SPACESHIP CONSPIRACY

GEORGE KNAP

GEORGE KNAP PUBLISHING

SPACESHIP CONSPIRACY

ISBN 0-9695767-0-6

KNAP PUBLISHING

4408 Wildwood Crescent
Burnaby, British Columbia
Canada V5G 2M4

Telephone (604) 433-7415

DEDICATED TO MY WIFE BEA

For her devotion throughout our married life
and who stood by me through thick and thin.

Without her assistance and encouragement,
this book could not have been written.

George Knap

CONTENTS

PROLOGUE

I was sitting alone in the living room of my home, assembling my thoughts and concentrating them on space propulsion, a subject that has always fascinated me and one that I always came back to again and again. I rose from my chair, turned and stared into the mirror above the fireplace, when my mind began to race. I do not know what triggered my thoughts that day, but immediately I could see, in my minds eye, an orbital space propulsion engine firing away. I had a mental picture of combining the rocket thrust with centrifugal forces. After thinking about this for a few minutes, I realized that I may have stumbled on something important. This sort of scared me because if my thought experiment of combining the rocket thrust with centrifugal forces did work, future space travel would be changed forever. I also realized that this would change the laws of physics. I was in awe of the concept myself.

The following story of orbital space propulsion should be told to the public as a prime example of things that inventors, with the exception of a chosen few, have to endure. Like a lot of dedicated inventors, I have found that the most difficult things in life have always appealed to me. The easy things I have found wanting and I have never feared to venture into the unknown.

THE BEGINNING

I was born in Vancouver, British Columbia, Canada on November 3, 1924.

My mother's name was Rachel, my father's name was Morris. I have two sisters, Sarah, who is older than me, and Sylvia who is younger.

The only place I can remember living in, as a child, was an apartment that was directly over a Japanese grocery store, named the Oma Brothers, which was located on the corner of Prior and Jackson Avenue in Vancouver's East End. My mother was a kind and loving soul who was a bit of a dreamer and who loved writing poetry. My father served in the Canadian Army in World War I, where he received a severe head injury caused by a falling piece of timber. After his discharge from the army my mother wrote hundreds of letters to the Canadian Government trying to obtain a pension for him. She did this over a period of years, but never did succeed. My father's head injury caused many erratic mood changes. He would treat me to the movies one time and beat the hell out of me the next time. I was often afraid to go home when he was in a foul mood and spent many nights sleeping in wood sheds and horse barns. We spent a lot of time living on relief (welfare) in the early days. My father was a junkman trying to augment relief as it was never enough to survive on. As a young lad I used to take a wagon, which I had built, and go around to various places picking up bottles, metals and anything of value. Most of the money I earned went to my mother. I believe I was around seven or eight years old at the time.

I will never forget the strange experiences I had at school one day. It was morning recess and I was playing in the school yard. It was a bright, sunny day. Suddenly, for no apparent reason, I looked up to the sky and saw what appeared to be a huge hole open up. It was awe inspiring, it was alternately smooth and rough, it was a happy feeling like I was in heaven with God. When I saw this the first time, I said to myself, I must remember this. The same thing happened again on the same day, only it was on my lunch hour, again in the school yard. It is something that I will never forget and I have always wondered what its significance would mean in my life. I believe I was also around seven years old at the time. My mother often told me that I was born with an aura about me and had always told me that I was destined for greatness. Everyone laughed at this, but at times she was psychic.

In my early years I loved reading, comics, fairytales and pirate stories. Later on I added history, geography and science of which I became very good at. What a wonderful thing the public library is. I still use it extensively today.

I ended up quitting school in grade 8, as I had to go to work as many others did in the depression days. After quitting school, I worked at odd jobs until the spring of 1939 when I went to work at the Empire Box Factory, located in the False Creek industrial area, now the site of B.C. Place Stadium. Expo 86 was held here also. The rate of pay was 25¢ an hour. This job was not challenging enough so I quit at the end of the year.

I was 15 in 1939 when World War II broke out. I had made up plans for ground to air guided missiles, which I submitted to the National Research Council in Ottawa. This was my first serious attempt at inventing, even though it was only on paper. I was pleased to receive a letter from General McNaughton, head of N.R.C. (National Research Council) commending me for my work at such an early age, but he assured me that the government was already working on this.

In the time period between 1940 and 1942, I was employed at three different shipyards. These shipyards were very busy. We were building freighters of ten thousand ton capacity to replace the losses of the freighters that had been sunk by German submarines.

One of my jobs at the shipyards was passing rivets. Each rivet gang composed of a riveter, his bucker up, a rivet heater and a rivet passer. The rivet heater would heat a number of rivets to a white heat in a coke fire. Using his tongs, he would select a rivet and throw it to a passer, who would catch it in a cone-shaped bucket. The rivet passer, using his tongs, would take a hot rivet from the bucket and place it in the rivet hole. The bucker up would use his pneumatic gun to hold the rivet in tight, while the riveters put a head on it. All this was done at a great speed before the rivet could cool down.

One of my other jobs was to collect and carry discarded rivets, nuts and bolts that were left over from the riveting crews. I also worked as a burner, cutting steel plate with an acetylene torch. I was also an engine fitter's (mechanic) assistant.

In 1941, when I was employed as a rivet passer at Burrard Drydock Co. situated in North Vancouver, I noticed piles of rivets, bolts, nuts, etc. all mixed together. The labour required to sort this material was not available. This material was accumulating at the rate of 50 tons per day. I used my time off the job trying to solve this problem. So being of an inventive mind, I built a sorting machine out of wood, scrap iron and auto parts and used a washing machine motor to power it. I built this machine in the family kitchen. My father thought the machine was a pile of junk and wanted to throw it out, but my mother wouldn't allow that to happen. Thank God for my mother!

When I finished building and testing the sorting machine, Melbourne Thomas, a friend of mine, brought his truck over and we loaded the machine on and took it to Pacific Drydock, but the foreman there was not interested.

We then took my sorting machine to Burrard Drydock Company in North Vancouver, B.C. Mr. Watson, foreman of the machine shop expressed interest and called in the president, Mr. Clarence Wallace to witness a demonstration. They were impressed, so much so that they decided to build a full production machine. I went to see Mr. Carter, Patent Attorney of Fetherstonhaugh & Company telling him that Burrard

Drydock was going to build a full production sorting machine and that I wanted a patent on it.

Over the next three months or so while the sorting machine was being built, Burrard Drydock gave me employment as a machinist's helper.

One day when I arrived at work, I noticed that the new sorting machine was gone. I found out that the machine was taken to Burrard Rivet & Forging Co. in North Vancouver. I then went to see the manager of the Burrard Rivet & Forging Co. and told him that I was the inventor of the sorting machine and that I had a patent on it. The manager said that he wanted to use the machine to sort out the material from Burrard Drydock. He then told me that I could operate the machine, which I did, for about three weeks at no pay. This was to be my last job working for others.

I never stayed long at any job where I had to work for others. I guess it's because I've worked and had done things on my own most of the time. I was destined to be self-employed.

I was then offered a contract to do all the yard's work for Burrard Drydock, as far as the sorting and rethreading of the nuts and bolts was concerned. On my next visit to Mr. Carter, my patent attorney, I was told that Burrard Drydock had telephoned him to ask if I had filed for a patent on my sorting machine. Mr. Carter replied that I had. One can only speculate as to why I received the contract from Burrard Drydock.

Mr. Carter sent me to Esmond Lando, a lawyer, to get financing to start up my own factory using my sorting machine. Instead Mr. Lando sent me to Pete Bingham of Bingham & Hobbs Equipment, to ask for financing. I had no luck with him.

At one time I had sold scrap for Moses and Julio Zanatta of Atlantic Transfer Co., so I decided to see them about financing. I offered them 25% each, while I was to get 50%, if they would supply the financing.They agreed.

We hired Esmond Lando to set up the Atlantic Sorting Co. Ltd. We rented a warehouse from Famous Foods located at 1306 Franklin Street in Vancouver. We also set of a line of credit to start the factory with the Canadian Bank of Commerce. Burrard Drydock supplied their threading machines plus the new sorting machine. We installed conveyor belts, manufactured another sorting machine and hired fifty workers. My sister Sally (Sarah) worked in the office along with Julio's wife and my sister Sylvia worked on the machine. In fact, several of my friends worked with me. We had three shifts going, twenty-four hours a day, seven days a week. The workers would pick out the rivets, bolts and nuts off the conveyer belts and the remainder, which was scrap metal, was dumped into bins at the end of the belt. The rivets and bolts went into the sorting machine. We did very well financially.

So at the age of seventeen/eighteen, I was half owner of a factory built around my invention of the rivet and bolt sorting machine. I must admit that gave me a swelled head for a time. We continued operating the plant until VE day, Germany's surrender. When this happened, immediately all wartime contracts were cancelled, closing the plant.

After Atlantic Sorting closed, I tried various other enterprises. Some were successful, others were not. I was always on a seesaw, it seems. I was in the scrap business, ship salvage, log salvage, plus numerous other enterprises. If I needed special equipment for a job that I was working on, I just designed something for that particular use.

My father was killed in 1947 when his truck ran off a steep long hill near the town of Merritt, B.C. after the brakes on his truck had failed. His driver escaped, receiving two broken legs when he jumped out of the moving truck. My father was on the side of the road that had a sheer drop. He had no time to escape. I guess that I cared for him more than I was willing to admit.

My mother died February of 1952, after a long battle with cancer. She was a great loss to all of us.

I met my future wife Beatrice on a blind date through friends in 1951. We were married on Easter Sunday, April 13, 1952. Our fortieth wedding anniversary is coming up shortly and we have been blessed with four beautiful daughters.

In the 1960's I was involved in the White Rock, B.C. Beach Development Project for which I invented and patented a submarine scraper. I also designed and invented a new type of steel rolling mill which I received a United States patent for.

INVENTION CONCEPTS 1950's

It has always been my lifetime ambition to become a very good professional inventor.

To this end, I have tried to blend my creative imagination with a very broad knowledge base. I have always believed that the whole key to this enterprise would be the flexibility of my mind. It is the norm that most people tend to form their thoughts around one or more comfortable ruts. I also realized that if I got into these thinking ruts, my mind would be robbed of the much needed flexible thoughts.

During my lifetime I have set aside, in my mind, hundreds of possibilities that would form the basis of my inventive ideas. This type of thinking would enable me to work on many inventive problems in the same time period. I assumed that by trying to solve a great number of these problems at the same time, I would increase the chances of expanding my mind as well as avoiding the pitfalls of stale thinking.

Among the different projects I have worked on in the past, were many inventions that I thought could be marketed in very large quantities enabling me to make money on sheer volume alone. I assumed that it would be easy to manufacture and market these items. How wrong I was!

The following is a list of some of the items I have worked on:

1. **Safety Safety Pin**
   This type of safety pin was designed in my early days of inventing. The idea was to avoid the danger of the pin point springing open and piercing the user.

2. **Hairpin**
   By squeezing the end of the hairpin, the pin would open enough to be inserted into the hair.

3. **Plastic Soft Drink Container**
   A soft drink plastic can to hold carbonated beverages which could also be used as a drinking glass.

4. **Twist-Off Bottle Cap**
   Carbonated beverage bottle caps would twist off without having to use a bottle opener.

5. **Jar Ring for Preserves (and ordinary glass jars)**
   Jar lids could be easily opened no matter how tight the jar is sealed by simply tapping on the raised portions of the lid.

6. **A Self-Cleaning Hair Comb**
   Dirt and hair would not accumulate between the teeth of the comb.

7. **Concrete Fence Post**
   Designed to screw into the ground without the necessity of digging a hole.

8. **A Tubular Plastic Container (with a tear-off strip)**
   This type of container is filled with different kinds of powdered beverages so that a person could immerse this flavouring tube into a glass or cup of cold or hot liquid.

9. **Beer and Soft Drink Beverage Cooler**

10. **Water Powered Rotary Car Washer Brush**

11. **Telephone Book Index**

12. **Barbed Wire (plastic barbs)**

13. **Nail Extractor (drill powered)**

14. **Potato Peeler (for household – powered)**

15. **Heat Exchanger**

This is a partial list of some of my earlier inventions.

ORBITAL SPACE PROPULSION 1968

Many years after I had invented the Rivet Sorting machine, I got in the habit of mentally setting up a list of the unresolved problems in the field of inventions and science.

Every so often I would concentrate on a particular problem from this list. Since I was fascinated with space propulsion, I devoted my thoughts toward this field.

On this particular occasion I was sitting alone in the living room of my home when I got up from the chair. I looked into the mirror directly in front of me and the thought that came into my mind was about combining two different forces into one force. Immediately I could see in my mind's eye a picture of an orbital propulsion engine firing away. After a few minutes thinking about this, I realized that there must be something to this idea. This sort of scared me because if this thought experiment of combining the rocket thrust with centrifugal forces did work, future space travel would be changed forever. Although I never had formal engineering training, I have always been able to form mental images of moving masses and forces. This ability extends inward from the smallest particle to the largest structure. Training myself, I have always looked at the overall concept of everything rather than all the minor details, because, if this concept did not work I would only be wasting time worrying about details. On the other hand, if the concept was feasible, I would go into detail. Doing things this way has made it amazingly simple for me to come up with original concepts. I suppose that over a long period of time I must have subconsciously trained my mind to accomplish this, so much so, that I felt like the old lady who lived in a shoe who had so many children that she didn't know what to do.

I don't know what triggered my thoughts when I looked into the mirror that day. I never dreamed of the problems I would have or the impossible situation it would put me in.

A few days later I got in touch with my friend, Bill Benz, who was a draftsman. Bill had done other work for me in the past. We discussed my concept of "orbital propulsion" to the point that he had enough understanding of it to make a drawing. I still possess Bill's original drawing.

After studying this drawing, looking for fault, I came to the conclusion that these propulsion engines must be operated in counter rotating pairs. The next thing I did was to conduct simple and inexpensive experiments to find out if the physics involved was correct. Some of my earliest experiments were done on a record player. I conducted these experiments by placing various objects on the rotating record at different speed levels. These simple experiments indicated that I was on the right track. I have always looked around in the physical world when I had a problem finding proof that would justify my theories. An example of this was that any person sitting on a swing would be able to gradually increase the arc of travel by changing their position at the beginning of the swing. The person on the swing could change their position by pulling on the two chains of the swing thereby shortening their length. Even a slight shortening of the length of the swings chains would cause an increase in the swings momentum.

I could not find any reason why my principal would not work. I also knew that, due to my limited finances and my lack of formal scientific training, this project would be too big for me to handle alone.

At that time, due to Canada not having a space program of its own, I approached the United States consul in Vancouver, British Columbia and arranged for a meeting. The U.S. consular suggested that the best place to send my material for "orbital propulsion" was to the United States Airforce. I took the precaution of having him stamp my original drawing with the U.S. consuls seal as well as the date. I gave the consular a photocopy of this drawing to send out to the U.S. Airforce. I received a letter from the United States Airforce Evaluation Department a short time later criticizing me for daring to refute one of Isaac Newton's laws that for every action there must be an equal opposite action. After studying this letter I was very upset but not discouraged. I still have this letter on file. In the meantime, I sent a telegram to the President of the National Research Council in Ottawa, Canada requesting assistance in this matter. I received a reply stating that my idea was impossible.

At the same time, I made arrangements to have a meeting with an ex-officer (Royal Air Force) who was head of the British Trade Commission in Vancouver to explain how my system worked. I gave him a copy of my drawing and the officer said he would send it to a government department in Great Britain.

In a few days I received a letter, addressed to me, from the British Trade Commission. Upon opening the letter, I read with great interest that this letter was supposed to be sent to the British government which more or less stated that I was a one man band and that "orbital propulsion" was a Heath Robertson devise (Americans call it Rube Goldberg). I was obviously sent the wrong letter. I was naturally peeved and telephoned this gentleman from the Trade Commission and asked for an explanation. He informed me that his secretary had mailed the letter to me by mistake and requested me to return the letter to him, which I refused to do.

I also tried to seek out scientific opinions from some of the professors at the University of British Columbia but to no avail. This is the beginning of a very long road.

SEEING IT THROUGH 1973

In 1973 I felt that I now had sufficient money available to carry on with my interest in the field of inventions. Therefore I decided to pick two of my invention projects and thoroughly test them so that I would have a sufficient amount of information to be able to apply for patents. One of my projects was a lumber slicing machine and the other project was my orbital space propulsion concept which has always been my first priority.

I have always been fascinated with the concept of orbital propulsion since that day in 1968 when I first got the idea in my living room and when I also knew that I had come upon a new law in physics. This type of orbital propulsion spaceship would be able to take off from any suitable airport, where it could fly directly to other planets or star systems at maximum speed and return to earth the same way. It's speed would only be limited by the laws of physics.

TESTING OF ORBITAL SPACE PROPULSION

The first thing I had to do in order to prove if my concept for orbital space propulsion would work was to design and build a working model.

It took me just under a month of intensive thinking to come up with a design for this machine that would be simple, flexible and economical to operate. After I put the rough design on paper, I went to see Herbie, owner of Ideal Iron Works that was located on Alexander Street in Vancouver. I have known Herbie for many years and have had a lot of business dealings with him. He gave the job of making up the machine to his son, who was a very fine machinist. After a week's time, the machine was ready for testing.

I had to have a test stand to secure the machine and to measure the thrust, so I went out and bought a fairly large dial scale. I took the machine and the scale over to our shop on Powell Street. Then I went next door to the Welding Supply Co. and made arrangements to purchase a large quantity of compressed nitrogen gas. This nitrogen gas was under the pressure of 2200 pounds per square inch, which was enclosed in large steel cylinders. First I tried testing with a single cylinder, but due to the gas pressure dropping too fast, I had a manifold made up complete with a gauge, so that I could use two cylinders at a time.

Many a time I got discouraged. Out of the many tests that we did, nothing seemed to be happening.

One day, when I started the machine up for a test, the four bronze nozzles went to the centre of the hub while the machine was rotating with the gas pressure on. Then the four nozzles would go outward against the four springs. I was quite amazed to see the rotor turning. I would guess it at a quarter turn and then it suddenly stopped even though the electric motor was developing full power. At the same time the intermittent motion was continuing, the machine sounded exactly like a machine gun being fired. I am sure, even though these tests were made inside the building, that the noise was heard a long way off. During this erratic behaviour, the machine jumped off the scale and hit the floor causing severe damage to the four rotor arms. It was then that I decided something was happening, although the results of this particular test was not what I was looking for. The four arms of the machine were made of stainless steel tubing and I could not use them again as they were bent, so I bought more stainless steel tubing to make new arms.

Leon, my business partner at that time, was also in the business of making stainless steel turnbuckles of the bottle type as well as swaging (shaping) stainless steel fittings for sailboat rigging, so I had Leon put one end of these tubes into his swaging machine to reduce the diameter of the tube on one end. After the four rotor arm tubes were machined, I decided to purchase four linear type ball bearing races. When I finished this, I installed the arms back on the machine. This drastic measure cut down the friction between the four bronze nozzles and the stainless steel rotor arms. I also limited the amount of travel to about one inch, as I did not want a repeat of the intermittent motion that caused the accident in the first place.

Another month had past and the machine, although running, was not producing positive results. All that it was giving me, in the tests, was a constant two and a half pounds of thrust as measured on the dial scale. This was due to the thrust of the escaping gas from the nozzle. The problem was that I could not see what was happening because the rotor arms were revolving too fast.

The next thing I did was to get in touch with Dr. Gygax, who was a professor in the Physics Department at Simon Fraser University in Burnaby, B.C. He took time out from his busy schedule to come down to the shop to look the machine over. He suggested using a strobe light to shine on the rotating arms of the rotor so that we would be able to see exactly what was happening. He said he could supply the strobe light which I readily accepted. A few days later he brought the light down with him. At dusk we were standing by the machine, in the dark. All the building lights were out. We could easily see that the four springs did maintain a certain frequency due to the effects of gravity upon the mass of the four nozzles. To get the machine to work properly I had to impose the full thrust of the compressed gas before I started the rotation.

In further tests, over the next few months, I managed to develop seven and a half pounds thrust, as measured by the scale, which is a ratio of approximately two to one. Two and a half pounds of this was due to the compressed gas thrust and five pounds was due to the continuous unbalanced centrifugal forces which was functional to the power input of the electric motor. All these tests were made with the exhaust gas shooting straight up into the air. When this machine was operating correctly you could easily tell that the small electric motor was under full load and the needle pointer on the scale would go up relatively slow until it reached a certain reading then it would remain there for several minutes without a quiver. Then as the gas pressure began to drop, the needle would go down to it's original position.

I can remember when I viewed this with some of my men, there was a lot of elation and enthusiasm and I drank it all in, savouring every minute of the first successful test.

By this time I had gone through two hundred cylinders of compressed nitrogen gas at the cost of $12 per cylinder, which lasted only about three minutes using two cylinders at a time. This was due to the rapid decrease in pressure.

I began to slow down the tests on the machine until I could find a scientist or engineer who could give me advice. The first thing I did was call up the British Columbia Research Council to ask for their assistance. They sent down an engineer to have a look at the machine. After looking everything over, he said that he couldn't figure it out, then left.

In the meantime I had further discussions with Dr. Gygax. He seemed to agree with me as to what was happening with the machine even though he hadn't witnessed a successful test. I next placed an ad in the student's newspaper at the University of British Columbia, where I advertised for a scientist or engineer to assist me as I wanted to be sure of my grounds before applying for a United States patent. I then received a telephone call, in answer to my ad, from a man who said he was an atomic physicist. I invited him over to my house for a meeting, where I found out he was an Egyptian in his thirties and who seemed intelligent. I agreed to pay him $200 per week to witness my testing and to help me understand the workings of the machine in technical terms. He never said very much, always agreeing with me and never giving me any information about the machine that I didn't already know. One day he arrived at my work place with an older gentlemen of sixty years or so. This gentlemen said that he was an engineering professor at the University of B.C. and that he did most of his work in Ottawa. I took this man at face value as he told me that he was Canada's leading expert in space science. After looking over the test machine, he said to me that just because the machine works on the ground, doesn't necessarily mean that it will work in space. I asked him what he meant by this as I thought that it was a stupid statement. I did realize then, that a great many rockets are tested in a static position to determine, among other things, the thrust capabilities of rocket engines. After a half an hour of discussion with this gentleman who was trying to impress me with his intelligence as well as trying to discourage me in my endeavors, I sort of got peeved with him and this ended the conversation.

Shortly afterward, I paid the Egyptian nuclear physicist the $400 that I owed him and that was the end of that. It was then that I decided to go ahead and apply for a United States patent on orbital propulsion.

Shortly after, I had a meeting with Mr. Jones, a patent agent with the firm Carver and Company. I had known Mr. Jones for five years or so. We had worked on other patents together. Mr. Jones was an ex-Canadian army officer, who had specialized training in the law as well as in engineering. I had always considered him a friend. After taking Mr. Jones to see the machine and trying to explain the principal involved, he seemed to grasp the significance of it. So that he wouldn't make any mistakes in the patent application, I advised Mr. Jones to get in touch with Dr. Gygax to advise him.

Later on, Mr. Jones made an application to the United States patent office for a patent.

DEEP SPACE LINER

I have also envisioned that someday the present type of space craft now in use will be superseded by very large deep space liners capable of making very long journeys to other star systems. This space vehicle will be propelled by multiple orbital propulsion engines. This type of space liner will utilize three separate energy sources: atomic, solar cells and chemical fuels. The chemical fuels will be mainly used to provide expendable mass. Both the atomic and solar energy sources will be used in the form of electricity to rotate the orbital propulsion engines.

Due to the huge size of this type of space liner, it would be practical for the liner to remain in orbit. Transportation from the planet's surface, for both man and materials, will be supplied by shuttle craft utilizing the orbital propulsion engines.

The orbital propulsion space liner's ability is to gradually accelerate its speed over a period of time, whereas the present rocket uses most of its fuel on the initial firing. Maximum speed is limited only by the laws of physics. Voyages can be taken in a more direct route thereby shortening the distance to its destination.

Orbital propulsion type space liners would be comparable in size to the largest ocean liner. Accommodations would be available for a thousand or more passengers and crew personnel.

This type of space liner would be built of lightweight material, constructed from prefabricated parts, and would be assembled in the earth's orbit.

By attaching many solar cells panels to the outside of the fuselage of this space liner, sufficient amounts of electric current could easily be generated to operate the main "orbital propulsion" engines and all other auxiliary purposes such as heating, lighting and the powering of electronic instruments on board. This generation of electric current by this method would provide a very low cost energy source for the space liner.

ORBITAL SPACE PROPULSION SHUTTLE

Space shuttles powered by orbital propulsion engines have many advantages over present day space shuttles. For instance, they are much more fuel efficient and have a greater degree of safety, both for its crew and the people who happen to be in the vicinity if a crash should occur. Due to this last, very important feature, this craft will have no separate need for special isolated launching sites as used in the present rocket propelled craft.

In addition, this orbital propulsion powered space shuttle will have the ability to operate like an airplane when it comes to take off and landing procedures on the surface of the earth. Matched pairs of counter rotating engines will supply the necessary thrust for lift-off and for braking its rate of descent back into the earth's atmosphere. Only chemical fuels will be used to power this space shuttle.

Also, owing to the orbital propulsion shuttle's ability to take-off from an ordinary runway at low speed and its gradual ascent through the atmosphere into space at a low rate of speed, some oversize cargo can be carried on the outside of the fuselage.

ORBITAL PROPULSION ENGINE PERFORMANCE SPECIFICATIONS

The following are specifications of orbital space propulsion engines:

A. Centrifugal rotational forces can be changed into a directional thrust for space propulsion.

B. By changing the distance of rotating mass from its axis, alters the rotational speed of the rotor. Travelling inward toward its axis speeds up the rotor, travelling outward slows the rotor down.

C. Rocket thruster nozzles represent the mass. Firing on regular pulse intervals, inward toward the axis of rotation in a given segment of the arc, changes the distance of the mass from its axis.

D. The firing of rocket thrusters accomplishes two distinct actions:
   1. Reaction of escaping exhaust from the rocket thrusters produces a propulsive effort upon the spaceship, propelling it through space in a normal way.
   2. On firing the rocket, the thruster nozzles is driven outwardly to a higher orbital position from its central axis to its furthest position. At the same time, by compressing the springs, it allows the rocket thruster to return to its original position when the firing pulse is completed and the springs return to normal.

   It is normal to expect, that by extending the rocket thruster nozzles outward into a greater orbit away from its central axis, would cause the rotor to abruptly slow down its rotation. The rotational speed remains constant due to the additional power input generated by the spaceship's main power source. By combining the main engine power and the thrust from the rocket thruster nozzles, the magnification of centrifugal forces is attained. These additional centrifugal forces will pull the rocket thruster nozzles into a still higher orbit from its central axis thereby creating an unbalanced centrifugal force in any given direction and harmonic motion is the direct result.

E. On the actual prototype machine test results, the ratio is two to one:
   1. For every pound of rocket thrust, two additional pounds is developed by the increase of centrifugal force. This ratio can be expanded to a higher value by increasing the rotor speed. The prototype machine on which these results were obtained was made to operate at a fixed rotational speed.

F. Orbital propulsion engines are operated in matched counter rotating pairs. This eliminates torque on the spaceship.

AT LAST – MY PATENT 1978

Over four years had past since my patent agent, Mr. Jones of Carver & Company, filed the first United States patent application for orbital propulsion.

The first application for the patent had a drawing of my test machine on it. I told Mr. Jones that the test machine was never meant to fly and that it was a prototype for testing purposes only. He never made that clear on the application form. This application was rejected by the primary patent examiner, Barry L. Kelmachter, because there was not enough information on it. On the second application for the patent, I hired an engineer to make more detailed drawings. Mr. Kelmachter rejected this application also. He was obviously trying to help me by making sure that everything was corrected. I was getting annoyed with Mr. Jones as each one of these new applications was costing me money.

Mr. Jones told me at one of our meetings that there was no way that I could prevent the United States Government from working on this invention and that they didn't need my permission even if I could get a patent. I got so angry that I told him that under no condition would I permanently abandon my quest for a United States patent and that if he didn't smarten up, I would initiate legal proceedings against him and his firm, Carver & Company. The problem seemed to be that Mr. Jones did or did not know what he was doing.

Even on the fourth and final patent application, he referred to my test machine rather than the engine itself. Even the name on the application was listed as "orbital propulsion apparatus" rather than just "orbital propulsion". Anyhow, after Mr. Kelmachter of the patented office assisted in the corrections of all the applications, I was finally granted a patent on May 2, 1978.

So finally after ten years, since I first had the concept of "orbital propulsion", I received my patent.

After I received my original patent on "orbital propulsion" plus several copies of the patents description from the United States patent office, I took one of the patent copy description papers to the printer and ordered an extra two thousand copies.

Then I began a mail out campaign in the effort to publicize my "orbital space propulsion" invention. These copies were sent to aero space companies, assorted magazines, etc.

I received some replies, most of whom suggested writing to someone else. As a whole, this mailing campaign did not produce the results that I had hoped for, even though I had kept this up for several years.

EXPO 86

In 1985, after learning that Vancouver, British Columbia was to host Expo 86 and that it's theme was going to be transportation and communications, I made telephone inquiries to the various Expo officials offering my orbital propulsion system as well as a few of my other inventions in the field of transportation for display. I offered this package deal for free, even though I only had a patent on orbital propulsion and not the others. I was not able to afford the costs of the patents on these other transportation inventions. I knew that I would lose my legal rights to them if they were publicized at Expo 86, or anywhere else for that matter. My main concern really was to publicize my orbital propulsion system and I was willing to sacrifice my other inventions to do so at the time.

After numerous attempts to arrange meetings with the Expo officials and failing, I approached the Imperial Oil Company Ltd., one of the Expo 86 sponsors. With their assistance a meeting was arranged with David Youngson, an official for Expo. At the first meeting he expressed interest in my work, but stated that my work must be presented in a more visual way. He gave me two weeks time to engage a suitable artist to have this work done, which was no easy task.

I hired a young, talented artist by the name of Glen Johnson who was referred to me by a friend and neighbour of mine, Jake Krahn. Glen made several drawings for me, which I presented to David Youngson on our second meeting. At this meeting with Mr. Youngson was a female executive, who was very enthusiastic about the drawings and also the fact that I was born in Vancouver and had actually worked on the Expo site itself when the Empire Box Factory was situated there years ago. Mr. Youngson stated that he wanted more detailed drawings which I agreed to have done. I even offered to pay for the entire display myself. He said that he would let me know whether or not my works would be displayed at Expo.

Glen agreed to make more detailed drawings on the orbital propulsion spaceship. I also had Glen make up a whole series of drawings in the field of transportation as well, illustrating my future invention concepts. One of these I named the "Land Tanker", which is a huge machine that could be used to transport fresh water or oil across arid regions where roads would not be required. This vehicle would have the potential to carry thousands of tons of liquid.

I met with David Youngson a third time, bringing him more detailed drawings on the orbital propulsion spaceship as well as a number of invention concepts, including the land tanker. He showed interest in everything except the orbital propulsion spaceship which I thought was odd. He also insisted, through all our meetings, that he would write me a letter, which he did, on everything that Expo wanted as well as what we discussed. I had several letters from him as a matter of fact. He would tell me in letter form why this or that was rejected and I would correct everything and so on. Once again, he told me that he would let me know whether or not my works would be displayed at Expo 86.

As time went on and the deadline for entry was approaching, I phoned David Youngson and asked him about displaying my work. He was very evasive, so I reached the conclusion of being strung along and I told him so.

Shortly after, while listening to CJOR Radio, Jim Pattison, President of Expo 86, was discussing the complaints from local would-be participants being denied access to Expo. Jim Pattison promised access to himself for those persons voicing complaints. It was for this reason that I arranged a meeting with Mr. Pattison at Expo and showed him my illustrations of orbital propulsion as well as my other futuristic inventions in the field of transportation. He was very receptive, telling me that of all the things he had done and all the money that he had made, none of it compared to the work that I had done on orbital propulsion and my other inventions. He told me that he was envious of my inventive mind. I was close to tears when Mr. Pattison told me this. I thought that I had finally found someone who believed in me and was also in a position to help me out. I was on cloud nine when I arrived home and told my wife Bea.

But this was not to last, for shortly after David Youngson phoned me, not at all pleased that I had gone over his head, telling me that he was interested in the land tanker only, leaving out orbital propulsion, which I considered my most important work. This ended up with me saying that if he would not accept orbital propulsion, I was not interested in displaying anything else at Expo 86. Why should I sacrifice my unpatented invention if I couldn't display orbital propulsion? All I wanted to display was a large artistic mural of future space travel using my orbital propulsion invention. I wasn't asking for a huge area to do this. When Expo 86 opened, my wife and I attended and all we saw were large banners on poles depicting various transportation themes. This is where mine would probably have gone. So you can see that I wasn't asking a lot of Expo 86.

This telephone call from David Youngson made me angry and upset. This made me decide to get in touch with Elwood Veitch who is the member of the legislative assembly in British Columbia for our district and a cabinet minister of the Social Credit Government in B.C. I telephoned him and asked if he could see me right away. He invited me to visit his office. At this meeting, which lasted under an hour, I showed him a number of letters that I had received from the officials at Expo 86. He appeared very interested and told me that he would have to leave that evening to return to Victoria, the Capital City of B.C. and the seat of the Provincial Government, and that he needed some of my letters to take with him so that he could make inquiries to the Minister responsible for Expo 86. As he was anxious to make his departure, I consented to loan him all the original letters that I had received from the officials at Expo that pertained to orbital space propulsion. He said that he would make copies of these original letters and would return the original letters to me immediately. I was grasping at straws, so what else could I have done? To this day, in spite of calling him by telephone at his office as well as his residence in Burnaby, I never did get back my original letters. All I did receive a few days after this meeting was a letter from the Social Credit Party inviting me to join them and asking for a cash donation.

Much later, I made a formal complaint to the Provincial Ombudsman requesting that they act for me in the effort to have my letters returned from Elwood Veitch.

After checking into my complaint, the Ombudsman wrote me a letter stating that all the correspondence relating to Expo 86 had been destroyed.

At the same time that I had made the complaint to the Ombudsman, I received a telephone call from Elwood Veitch's special assistant telling me that my orbital space propulsion invention had been expropriated by the United States Government for their Star Wars Program!

I wrote a letter to David Youngson making a final appeal, expressing the concern of his choice of my land tanker over my orbital propulsion to be displayed at Expo 86. I sent a copy of this letter to Mr. Pattison as well, thanking him for his assistance. I never received a reply from David Youngson or Jim Pattison.

I also approached the Canadian Pacific Airlines who had an office in the Canada building at Expo. When I had a conversation with an executive there, I was told that Expo 86 was a fun thing and that they did not want to get involved in anything controversial. Others in the Canada Pavilion refused to look at my work also because of their knowledge of my problems with David Youngson.

I then arranged a meeting with Tom Siddon's (former Science Minister) personal assistant, in his Vancouver office in a last ditch effort to overcome this problem I was assured by his personal assistant that they would back me at the Canada Pavilion. She knew in advance what my problems were. After waiting a time and not hearing from them, I telephoned to find out as to what was happening. She informed me that I would receive no help from this government department and that Mr. Tom Siddon refused to see me. This ended all hopes of my displaying my invention of orbital propulsion at Expo 86.

THE U.S.S.R. PAVILION AT EXPO 86

It seems that the more I am rebuffed, the harder I try, so it was with great interest that I watched a BCTV Television interview with one of the leading Soviet space experts, who was permanently stationed at the Soviet Pavilion at Expo 86.

As I had a seasons pass for Expo 86, I took the chance of meeting him without telephoning him first. I was curious to find out whether or not they knew anything of my orbital space propulsion. The U.S.S.R. had a very large pavilion at Expo and most of their exhibits pertained to past Soviet space programs.

As I entered their building, I went straight to the office and made inquiries of this gentleman, asking to see him. I must admit that I cannot remember his name now. I handed the female Russian secretary a copy of my orbital space propulsion patent. The secretary located this gentleman and showed him my patent copy. This space expert came back to the office accompanied by another man and a woman. The space expert spoke excellent English and they ushered me out of the office and took me to a remote part of their building by one of the walls. I guess he thought the office was bugged. I noticed that the other man and the woman did not seem to understand English very well. This meeting went on for an hour or so with me explaining how orbital propulsion works and with him translating to the other two in Russian. One thing he did tell me was that they had no knowledge of my orbital propulsion in the U.S.S.R. All in all another meeting was suggested at a later date.

It was my feeling that they did not seem to trust one another. They seemed afraid to say what they thought. I had the impression that all the people in the pavilion were afraid to say what they thought. This surprised me because they were so unlike the Chinese, who stuck together. Nothing more was said or done at this point.

ME A SPY? NO WAY

After being turned down in my efforts to display my orbital space propulsion discovery in the various pavilions at Expo 86, I became extremely frustrated, so much so, that I decided to get an expert opinion as to what was going on, from the Chinese Consulate located at 3380 Granville Street in Vancouver, B.C. So I telephoned the Consulate and arranged a meeting with Mao Weiming, the Vice-Consul. At the consulate I was greeted by Mr. Weiming, who initially showed a considerable reserve, but after showing him my United States patent for orbital space propulsion and some of my other inventions, his interest became enthusiastic, so much so, that he introduced me to the consular, Mr. Jiang Qifa, and other members of his staff. As we sat around the table drinking tea, they had an animated discussion in the Chinese language. I gathered that their main interest was the patent on orbital space propulsion, which was examined closely.

After finishing the preliminary examination of my patent, Mr. Weiming told me that he believed my invention had been stolen by the United States Government because of the seven year time lapse since my patent was first issued by the United States patent office. The only comments I made to him about this seven year lapse was to say that I had tried various ways to publicize this invention and failed, even on my efforts to have it displayed at Expo 86.

This first meeting with Mr. Weiming lasted for around two hours and ended up with him proposing that I submit a resume to the Consular, Mr. Qifa. Most of my meetings were with Mr. Weiming, as he was the only one who spoke excellent English. He told me the reason for the resume was that they were interested in sending me to China to organize a special school to teach the Chinese students the process of making inventions. I thought to myself at that time, at last someone was interested in my inventive works.

I decided to discuss this matter with my wife Bea when I arrived home. Bea and I had a heated discussion about going to China to live. She didn't like the idea at all and was suspicious of the Consulate's motives. She told me that no way she was going to uproot the family, but that if I wanted to go that badly, I could go myself. She also warned me that if I did go to China, there was a possibility that they wouldn't let me leave the country.

I must admit that I was sorely tempted to go myself. So much so, that I had a resume made up and delivered it to Mr. Weiming a few days later. The resume is as follows:

AUGUST 22, 1985

JIANG QIFA
CONSUL OF THE CONSULATE GENERAL
OF THE PEOPLE'S REPUBLIC OF CHINA
IN VANCOUVER
3380 GRANVILLE STREET
VANCOUVER, B.C. V6H 3K3

Dear Sir:

The following is a brief history of my accomplishments in the field of business and industry.

I was born in Vancouver, B.C. on November 3, 1924. As my parents were poor, my education consisted of the 8th grade at Strathcona School in Vancouver.

On leaving school, I went to work in a box factory at the beginning of World War II. I found employment in the ship building yards. As a result of my employment at the shipyards, I noticed a vast accumulation of nuts, bolts & rivets which was used in the bolting up of plates prior to riveting. As these plates were riveted together, the bolts holding the plates were discarded, gathered up and put in large piles.

I then designed and invented special machinery (which I built in the kitchen of my parents apartment). On completion of this crude sorting machine, I had it transferred to North Vancouver to a company called Burrard Drydock whereby I demonstrated the merits of this machine. The manager of the shipyards agreed with these merits and ordered the machine shop section to construct, along the same principals, a full size working model. On completion of this machine, they offered me a contract plus additional threading and tapping machines and to receive from them, all these nuts, bolts etc., to process it and deliver it back to the shipyards in the condition they could use again.

As I did not have the money to finance this undertaking, I made a business deal with two brothers to finance the construction of this factory on a 50/50 basis. In order to do this work, I had to design and fabricate many types of new machinery and also act as general manager in charge of operations, employing up to 50 people.

This went on for a number of years, until I realized the war was near it's end. I then began another factory to manufacture Civilian goods, that is, metal stampings and aluminum castings.

After the war, I developed processes to melt down wartime aircraft and manufacture aluminum pots and pans out of the metal. Owing to government regulated price fixing, I could not sell the goods this plant manufactured, even though there was a good market for these goods, so I had to close the plant down.

Around 1947, I purchased a landing craft from the government and went into the marine salvage business. After a number of years, I sold this business and went into the trucking and trading business, buying and selling surplus equipment and scrap metals.

This was followed by the shipbreaking and conversion business, mainly cutting up ships for scrap and making floating restaurants and hotels out of some of these vessels.

After this I decided to buy two- fifteen hundred ton capacity barges mainly to transport scrap iron from upcoast points to the United States. I did this for two years, then changed over to hauling, with these vessels, low grade and sunken logs.

Finding this business not profitable, I went into the new and used machinery business. Bear in mind that in all these businesses, I used novel methods and inventions to do same.

For some reason or another, I have always taken the unsafe and unsure way. To make a long story short, I have developed a great number of new inventions, which after spending a great deal of time, money and effort, I find no success in marketing the same. Except for the inventions I have used myself for business purposes, I have not received one dollar.

For the past five or six years, I have been importing goods from overseas and selling same, mainly at swap meets, which is where I am today.

This meeting was on a friendly basis and before I left, he said that he would make inquiries about the school project in China and would be in touch with me.

The following year, in early Spring, the Expo 86 opened. I did not see my orbital space propulsion on display (I still hoped that by some miracle that it would be there).

This made me angry so I decided to get in touch with Mr. Weiming at the Chinese Consulate to find out what was happening regarding this special school in China to teach students the process of making inventions. Mr. Weiming invited me to the consulate for a meeting. Upon my arrival at the consulate, he ushered me into a special room. This room was on the main floor. It was a long and narrow room much like a large closet. There were no windows, and the only thing in it was a few chairs. Mr. Weiming got down to business and asked me a lot of questions. At that time, I was so angry and frustrated with the Canadian and the United States Government for not helping me that I offered him my invention and invention concepts on an exclusive basis.

It was then that Mr. Weiming told me that they knew that I didn't have an office in the Vancouver vicinity, but they would make arrangements for me to acquire a suitable office and a business to go with it, if I would be interested in acquiring information for them on a regular basis. It was then that it dawned on me that this was a proposal for me to act as their representative in the spying business.

I immediately got uptight at this suggestion and told him that I loved my country, Canada, and although I was angry and frustrated with her at the moment, I would not act for them. I said that I liked China but I loved Canada. After my outburst of shock, Mr. Weiming left the room. I waited for 10 to 15 minutes for him to return. When he didn't return, I left the room, went out to my car and drove home. Even on my drive home, I was extremely agitated by the turn of events. It finally dawned on me that they were only interested in recruiting me as a spy all along.

Upon arriving home, I decided that the best thing for me to do was to report this information to the appropriate Canadian Government Agency. I telephoned the office of the Canadian Intelligence and Security Services, giving them a brief outline of what had happened. They took my name, address and phone number and said that a representative would be over to my house shortly. Needless to say, no one from this office came around to discuss the events with me. The next day, I telephoned them again and they told me in a round about way, that they already knew all about me and that they did not want to discuss the matter with me. So I came to the conclusion that something very fishy was going on in regards to my orbital space propulsion and the Canadian Government.

PETITION

Shortly after my meeting with the officials of the U.S.S.R. at Expo 86, I decided to send out a petition, in the form of a long letter, which would give a general outline of my difficulties in getting publicity for my discovery of orbital space propulsion.

With the help of my wife Bea, these letters of petition were, at first, sent to *all* the members of the Canadian House of Commons and the Canadian Senate. These letters were mailed on or near September 17, 1986.

A short time later, we changed the heading on these petition letters and mailed them out to *all* the members of the United States Senate and the United States Congress. We sent out hundreds and hundreds of letters in this time period.

I received a reply from the Canadian Minister of National Defense which was short and meaningless.

I also received a reply from the Canadian Minister of Science and Technology stating that the laws of nature and physics was not patentable.

I already knew this. I want recognition for the discovery of this new law in physics, as I have stated, over and over again, as well as recognition for my patent on orbital space propulsion.

In this same time period, I received a long distance telephone call from the Secretary for the Leader of the Opposition., the Canadian House of Commons, Mr. John Turner, who wanted to know what was going on. I filled her in on the details, but I never did receive a reply from them, written or otherwise.

Around this time I also phoned Svend Robinson, a Member of Parliament from my own riding in Burnaby, B.C. I had a meeting with him in his Burnaby office on Kingsway. At that meeting I had given him a lot of written information along with my patent copy of orbital space propulsion. I then asked him if he would support my cause for recognition on the floor of the House of Commons by asking pertinent questions to the appropriate cabinet ministers. He did not give me a definite answer to my request, but said that he would make copies of my material and make sure that it was distributed to all the 282 members of Parliament. After a month's time had gone by, I still never received any replies from any member of the House of Commons, so I decided to telephone all the Members of Parliament at their constituency offices, who were from the Vancouver and Fraser Valley area, to find out if they had received or had knowledge of the papers that Svend Robinson had sent out. Each and everyone of these members told me that they had no knowledge of orbital propulsion and never received any papers regarding it. So naturally I came to the conclusion that none of the

Members of Parliament received this information from Svend Robinson.

So, with Bea's help, I decided to mail the same information package that I had given to Svend Robinson to all the Members of Parliament myself. One thing to my advantage was that this mail could be sent postage free. Shortly after mailing these letters, I received a phone call from Mr. Robinson in Ottawa, Ontario. I could tell by the tone of his voice that he was upset and he said that he did pass out all the written information that I had given him, to all the Members of Parliament. I told him, in no uncertain terms, that not one of the local Members of Parliament knew anything of my difficulties regarding orbital propulsion and this was the reason that I mailed the information to all the M.P.'s myself.

Later on, I telephoned him at his office in Burnaby. This was when he informed me that he would not get involved in pursuing this matter in the House of Commons. That ended our meeting.

A few weeks later, I received an invitation from Svend Robinson's office, to attend his Christmas Party. So once again, I met him at his office where I formally requested a letter from him, stating that he would not act for me in regards to orbital space propulsion. Mr. Robinson refused to do this. Therefore I told him, that if he did not give me this letter, I would attend his Christmas party and tell his guests of his unwillingness to help me. Finally he agreed to give me a letter stating that he would not act for me. The letter is as follows:

December 12, 1986

Dear Mr. Knap:

I recently received the information you forwarded to me concerning orbital propulsion. At your request, I made copies of the documents and circulated them to all 282 Members of Parliament. I met with you in my office on this issue. I do not feel that another meeting would be productive, and am thus not prepared to meet with you on this issue, or to raise the question of orbital propulsion in the House.

If I can be of assistance on any other issue, please do not hesitate to call me.

Sincerely,

Svend J. Robinson, M.P.
Burnaby

I received a better response from the senators and congressmen in Washington, D.C., U.S.A. than I did from my own Canadian Government. These senators and congressmen from the United States Government more or less expressed sympathy for me. I believe that as a direct result of my mailing campaign to them, the top officials of NASA were forced to send me a meaningful reply in the form of a letter dated March 12, 1987. The letter is as follows:

Dear Mr. Knap:

Your letter of September 17, 1986 to Dr. Harris was referred to our office. I apologize for the delay in answering your inquiry, but we wanted to be sure that we could provide you with a meaningful response.

It is evident that you have spent a lot of time and effort in developing your orbital propulsion apparatus concept. It is surprising that you haven't gotten more outside interest, or at least some answers to your questions.

Our assessment of your concept finds no fault with the mechanical design and agrees that indeed forward thrust would be achieved and that acceleration would be imparted to the spacecraft, but the question is: How much? In our judgement, so much inert mass has been added (nuclear power system and complex mechanical devices) that no matter how efficiently the resulting thrust is produced, the overall effect on spacecraft acceleration is probably less than if it were powered by a conventional propulsion device.

A more precise assessment is not feasible without a calculation of the total propulsion system mass and, equally important, a calculation of how much useful thrust is actually developed per expenditure of rocket propellant during each revolution of the rotating arms.

In the area of advanced propulsion, which is our business, whenever we propose a new concept, the first question asked is: How much better is it than the systems we have today? In the case of your unique design, you need to have such a comparison made in order to prove (or disapprove) your case.

Thank you for taking the time to share your ideas with us.

Sincerely

Edward A. Gabris, Director
Propulsion, Power and Energy Division
Office of Aeronautics and Space Technology
National Aeronautics and Space Administration (NASA)
Washington, D.C.

The following is my reply to this letter:

March 25, 1987

Dear Mr. Gabris:

Thank you for your letter of March 12, 1987. In paragraph three of your letter regarding your question on how much?

The answer is:

The test I made years ago on my prototype machine produced a clear and steady thrust ratio of two to one. (i.e.) For every pound of rocket thrust, I got two pounds of additional thrust. Whether or not this thrust ratio can be improved upon remains to be seen.

Due to the supposedly lack of interest and no knowledge feedback to me from any source, including NASA for almost 20 years, I found it impossible to continue this work without support.

As outlined in my letter of September 17, 1986, I separated the two categories, namely; space shuttle and space liner. As I visualized my invention for space shuttle, this craft would have an aircraft type take off and landing and would be entirely self-contained.

At the present time, I do not visualize an atomic type power source to supply energy to rotate the rocket nozzles. It is my belief that a non-atomic aircraft type prime mover can be utilized to power this craft. It is also my belief that direct rocket propulsion is extremely inefficient when utilized as a mass mover from earth to space.

I also believe that in the transportation of human beings into space, safety and reliability is the prime consideration.

As for the deep space liner type spacecraft, it would have to be assembled in space and it would be powered by three power sources; atomic energy, solar energy and chemical fuel. As this type of spacecraft would be permanently utilized in space, size and mass would not be limited as in the case of the space shuttle.

I wish to emphasize again, my orbital propulsion test machine has always been available to any organization which would take the trouble to examine it.

It has been my lifetime belief that often times there is more than one way of overcoming problems in every field of human endeavor.

I also have a concept for a new type of earth to space transportation system that has nothing to do with orbital propulsion, but, would seem to me to be entirely satisfactory for moving non-human cargo into space at a very low cost.

I remain,

Yours Truly,

George Knap

Please note that I never did receive a reply to this letter.

THE ACCESS TO INFORMATION AND PRIVACY ACT

I decided to make applications to the various Canadian Government Departments under the access to Information Act to find out if they had any information on file regarding orbital propulsion and myself and if it had been banned from publication in Canada under the Official Secrets Act.

I mailed my first application on February 20, 1989 to the Privacy Commissioner of Canada. I received a reply from them on March 15, 1989 referring me to the Department of Consumer and Corporate Affairs or the Justice Department in Ottawa. I also sent the request for access to information to the Science Council of Canada, who replied on April 14, 1989, stating that there was no material regarding orbital propulsion in their files. I also sent a letter to the Department of Industry, Science and Technology in Ottawa. I received a reply from them on April 18, 1989 stating that they had no record regarding orbital propulsion and referred me to the Information Commissioner. I received a reply from the Department of Justice on April 19, 1989 stating that their search revealed nothing relevant to my request, but that I should be hearing from one of the departments in due course, where I had submitted similar requests. I had also written to the National Research Council of Canada and received a reply from them on April 20, 1989 stating that they had four letters in my file, which I will be adding to the reference section at the end of the book. I also sent a letter to the Consumer and Corporate Affairs Department and received a reply from them on April 26, 1989 stating that patents are excluded from the access to Information Act. I have not got a Canadian patent on orbital propulsion. In reply to a letter that I had also sent to the National Defense Headquarters, they stated that I should get in touch with the Information Commissioner. They had also enclosed a couple of copies of notes made on the papers. I received a letter from the Privy Council of Canada on May 2, 1989 stating that there were no records of myself or orbital propulsion. I sent the same request to the Royal Canadian Mounted Police and received a reply from them on April 28, 1989 stating that no information pertaining to orbital propulsion was located and referred me to the Information Commissioner. I sent a letter to the Canadian Security Intelligence Service (CSIS) and received a reply from them on May 18, 1989 stating that a search of their records concerning orbital propulsion has been made, with negative results. I received a letter from the Department of Communications on May 10, 1989, which I am going to quote below:

May 10, 1989

Access to Information and Privacy Secretariat

Dear Mr. Knap:

This is further to your access to information request of April 24, which was received in this department on April 27, 1989.

A search of our files reveals that no records exist pertaining to orbital propulsion and the U.S. Patent No. 4,087,064. For this reason we are returning your cheque of five dollars covering the application fee.

Should you require further assistance, please do not hesitate to call me at (613) 990-0403.

Yours Sincerely,

Stephanie Perrin,
Coordinator,
Access to Information and
Privacy Secretariat

I telephoned the number in this letter and asked for the person who signed it. I was switched to several different persons before I was finally talking to a woman with a French Canadian accent who seemed to know all about me and orbital propulsion. She told me that the Canadian Government did not want the wrong people to have any information regarding orbital propulsion. I was so stunned by this admission that I couldn't promptly reply to her statement. Finally I said to her, don't you think that I should be paid for my work, even if this was so. She replied in an affirmative manner, but didn't offer any suggestions as to what I could do about it. To this day, I do not know if this was a cruel joke being played on me or if it was just a discreet warning.

After thinking this situation over, I decided to find a lawyer who would act for me on my appeal to the Commissioner of the Access to Information Act, and if this failed, my last resort would be to take the case to the Federal Court of Canada. The one thing that I have learned in dealing with the courts in the past is that I would have a better chance if I hired a lawyer who practices criminal law, as he usually has complete loyalty to his client, with no entanglements, more so than a corporation lawyer.

The following letter is an example of the obstacles I have had to face in trying to find a suitable lawyer.

May 18, 1989

BOLTON & MULDOON

BARRISTERS & SOLICITORS

Dear Sir:

RE: Retaining Counsel in Respect of Obtaining Information on Orbital Propulsion.

I confirm our telephone conversation of May 4, 1989. In that conversation, you informed me that you have demanded from the Federal Government of Canada information with respect to orbital propulsion, a process or mechanism of which you claim to be the inventor. Further, I understand that you are not being provided with the information that you seek and anticipate further difficulty in being provided with that information.

From our conversation, I understand your need for the information to be two fold. First, it seems that it is your understanding that your invention is being used or considered for use by the Federal Government of Canada and you are not receiving any credit whatsoever for your invention. You informed me that by credit you mean that you are neither receiving any monetary compensation nor recognition. Secondly, I understand that you feel that you are unable to publicize your invention and that the Federal Government may be suppressing your efforts to do so.

I further confirm that you informed me that you were dissatisfied with counsel with corporate and commercial experience. Further, I understand that upon your reading of "Civil Rights" a book authored by P. Michael Bolton, you thought that your difficulties would be rest resolved by retaining counsel with criminal experience.

Mr. Bolton and myself have given your problem careful consideration. Both of us have a sense of the frustration that you feel in respect to this problem. However, it is our collective opinion that the problem that you now face lies within the area of law known as intellectual property. As you no doubt know, intellectual property is the area of the law dealing with patents, copyright and trademarks. As criminal lawyers, we have no experience in this particular area of the law. It is our opinion that we are not sufficiently versed in this particular area of the law to adequately represent you in this matter. Further, it is our opinion that we would be doing you a disservice if we were to represent you.

It is our suggestion that you retain counsel with some expertise in the area of intellectual property. It is possible that by canvassing law firms with such expertise, you may find someone with the qualities that you desire. In your search, you may find someone that has the expertise in the area of intellectual property and in the area of criminal law.

Enclosed are two photocopies from a legal directory indicating firms that have expertise in intellectual property. Because intellectual property is an area of law which which we have no regular contact, we cannot provide you with a personal referral. It is hoped that the names provided will give you a start in your search for appropriate counsel. As well, I have enclosed those materials with respect to orbital propulsion which you provided to Mr. Bolton when you last met with him.

I am sorry that we can help you no further. However, I am grateful to you for considering our firm for this matter.

Yours truly,
Bolton & Muldoon
Per: Michael Klein

I was in contact with the firms on the list of lawyers whose expertise was in intellectual property, provided by the firm of Bolton & Muldoon. Not one of them however, had any expertise in criminal law as well. They were all in the corporate field. So in this endeavor I failed.

On January 9, 1989, I decided to write a letter to the Right Honourable Brian Mulroney, the Prime Minister of Canada.

The letter is as follows:

January 9, 1989

Dear Sir:

I appeal to you on behalf of the advancement of science and technology and myself to assist me in the publication of my discovery of orbital propulsion, United States Patent Number 4,087,064.

I sincerely believe that despite many, many attempts to publicize this important information, both in Canada and the United States, my efforts have been thwarted.

Enclosed is additional information, including a letter from the chief of propulsion from NASA, stating that my discovery works, but has doubts as to its efficiency. If this is so, why am I not recognized for discovering a very important new law in physics? Also enclosed is a letter from my M.P., Svend Robinson, refusing to help me in this matter.

I am requesting you to act as my member in the House of Commons regarding this matter. If you do not accept this nomination you will leave me disenfranchised.

Looking ahead in time, what would future generations of Canadians think of a Government who would not assist one of their own?

Respectfully Yours,
George Knap

I received the following reply to this letter from the Prime Minister's office on April 26, 1989:

Dear Mr. Knap:

On behalf of the Right Honourable Brian Mulroney, I would like to acknowledge receipt of your correspondence of January 9 requesting assistance in the publication of your discovery of orbital propulsion. I apologize for the delay in replying.

The Prime Minister appreciates your taking the time to write and inform him of your views. I have been asked to forward a copy of your correspondence to the Honourable Harvie Andre, Minister of Regional Industrial Expansion and Minister of State for Science and Technology, and the Honourable Bernard Valcourt, Minister of Consumer and Corporate Affairs. Please be assured that your comments will be given every consideration.

Thank you for writing the Prime Minister.

Yours Sincerely,

William Jory
Special Assistant

The following letters are the ones I eventually received from Harvey Andre, Minister of Regional Industrial Expansion:

June 13, 1989

Dear Mr. Knap:

The Prime Minister's office has forwarded to me a copy of your correspondence dated January 9, 1989 in which you expressed concern about the status of your patented orbital propulsion apparatus. You may consider this letter as an interim reply as I have requested the National Research Council (NRC) to look into the matter for me. I will write to you again when I have more information from NRC.

Yours Sincerely,
Harvie Andre

June 29, 1989

Dear Mr. Knap:

Further to my letter to you of June 13, 1989 in which you expressed concern about the status of your patented orbital propulsion apparatus, I have now received the following information from the Canadian Space Agency on the material you provided.

The propulsion effect is due only to the thrust of each rocket as it is fired. This is most easily explained in terms of the law of "Conservation of Momentum". The only momentum leaving the spacecraft is the exhaust gas from the rocket, and hence an equal and oppositely-directed momentum is gained by the spacecraft.

Your discussion of centrifugal force neglects the fact that, if the rotor is unbalanced, then the spacecraft wobbles, producing an opposing centrifugal force. It is similar to a man doing the hammer throw in the Olympics. Both man and hammer rotate about an axis between them rather than the hammer rotating about the man.

In view of the first comment, the proposed system is probably less efficient than a simple rocket.

I hope that you will continue in your research endeavors and I wish you every success.

Yours Sincerely,

Harvie Andre

It seems to me that all along a game is being played and that the focus of this game is to minimize the importance of my orbital propulsion discovery. I believe, as well, that I have been a victim of a well-organized conspiracy to prevent me from publicizing this invention in the media. Despite all my attempts to obtain professional help to determine the *equations of motion* for orbital space propulsion, I could not find any organization, physicist, mathematician or engineer who was willing to assist me, for love or money.

Most of them never referred to the letter from NASA and those that did obviously didn't grasp the significance of it. I assumed that they either didn't understand it or didn't want to understand it.

After reading the last letter from Harvie Andre dated June 13, 1989, I came to the conclusion that it would not have mattered what reply I made to him, it would have been in vain. The only chance I had was to find one or two organizations with a lot of prestige that might be able to help me overcome the objections in his letter. So accordingly, I decided to approach the British Columbia Research Corporation, which shares the same premises with the National Research Council, situated on the grounds of the University of British Columbia.

As a backup, I approached the officials of MacDonald Dettwiler, a large aerospace company situated in Richmond, B.C.

The following statement is taken from my daily journal at the time the events were happening:

Thursday, July 13, 1989

First meeting with Bill McEachern of the B.C. Research Corporation. I took my orbital propulsion test machine to show him at this meeting. Discussed with him the possibility of them providing the calculations for the equations of motions for orbital propulsion.

I left the meeting, arrived home, received a phone call from Mr. McEachern to the effect that they would require a payment of $5,000.00 to do this work. He also stated that part of this work would be paid by the British Columbia Technology Assistance Program (TAP). I told him, that at this time, all I could afford to pay was $2,500.00 and if the TAP program would match it with $2,500.00, it would bring the total to $5,000.00. We agreed to the arrangement in this same phone call. From this phone call onward, he was no longer returning my calls as to when the work was going to begin. Finally, I phoned his superior Dr. Malcolm Smith and arranged a meeting with Bill McEachern, Dr. Smith and myself for 9 a.m., August 10, 1989.

Friday, July 14, 1989

Had a meeting with Peter George, chief engineer at MacDonald Dettwiler and five other persons on his staff at their Richmond office. I took my test machine to show them as well. They appeared to be extremely interested in orbital propulsion and agreed to try to find the equations of motion. They offered to do this for me at no cost. I was delighted, of course. They told me that this information would not be ready until September 1989.

Thursday, August 10, 1989

Meeting at B.C. Research Corporation at 9 a.m. with Dr. Smith, Bill McEachern and myself. The purpose of this meeting was to find out why my phone calls to Bill, regarding my work, were being ignored. Dr. Smith is Bill's superior and was there at my request. Bill apologized for ignoring my phone calls, so I gave him a cheque for

$2,500.00 subject to the TAP program matching the $2,500.00. This $5,000.00 was to be used for calculations to get the equations of motions for orbital propulsion.

Tuesday, October 17, 1989

Telephoned Bill at B.C. Research to inquire about the progress being made in the development for the equations of motion for orbital propulsion. As the TAP grant was in place and my share paid, I was anxious to have this done. But Bill stated that he wanted a total of $25,000.00 to do the job and that they wouldn't begin the work until they had this money. I felt like I was being played for a sucker, so I told him that I wanted to have my $2,500.000 returned to me. He informed me that he would have Dr. Smith phone me the following day with the answer.

After this conversation, I phoned Peter George of MacDonald Dettwiler, a space research company. It had been two months since I last phoned him and I wanted to find out if any progress had been made on the equations of motions for orbital propulsion. He told me that they did not have the expertise to do the equations and suggested that I get in touch with the B.C.I.T. Physics Department to see if they would be interested in having this done as a class project. So I phoned Dr. Donna McDuff, head of the Physics Department at B.C.I.T. (British Columbia Institute of Technology), and arranged a meeting for October 20, 1989.

Wednesday, October 18, 1989

Extremely upset with the conversation between myself and Bill McEachern of B.C. Research.

Phoned the Association of Professional Engineers to find out the qualifications of Bill McEachern and was told that he was *not* registered with their association and was advised to get in touch with Mr. Lambert of Professional Engineers, who was away, but would be back on October 24, 1989. Phoned the Ministry of Advanced Education and Job Training. Talked to the head of the department, Gwen Upton, phone 660-6268. Made inquiries about the B.C. Provincial Grants in regard to B.C. Research Corporation. She replied that she never heard of the TAP program when I mentioned it. She said that she would phone Victoria, B.C. and get back to me the same day. I phoned Gwen Upton before she got in touch with me and she told me that she found out that the TAP program was a special arrangement that came directly out of Elwood Veitch' office. The reason I phoned Gwen was to cancel the B.C. Research' $2,500.00 grant.

Thursday, October 19, 1989

Phoned Dr. Terry Howard, president of B.C. Research Corporation requesting the return of my $2,500.00 which was to be used for the equations of motion for orbital propulsion. He stated that he knew nothing of this work but would check it out and get in touch with me on Friday, October 20th. He never phoned back.

Phoned Catherine Denhoff, attorney for Davis & Company. Received an estimate of $100.00 to send a letter to B.C. Research demanding the return of my $2,500.00.

Friday, October 20, 1989

Sick with the flu today.

Cancelled 10:30 a.m. appointment with Dr. Donna McDuff, Physics Department, B.C.I.T. Re: orbital propulsion.

Monday, October 23, 1989

Received, by mail, Invoice Number 49776 from B.C. Research Corporation.

Tuesday, October 24, 1989

Phoned Mr. Lambert, head of B.C. Professional Engineer Association, regarding the problems with B.C. Research Corporation.

Phoned Dr. Terry Howard, President of B.C. Research, to find out what he was going to do about the refund of my $2,500.00. He stated that if I would agree to sign a letter that he was going to mail to me, I would get a full refund. I gathered from this conversation that this letter would absolve B.C. Research of all liability pertaining to the calculations for the equations of motion for orbital propulsion.

Wednesday, October 25, 1989

Delivered envelopes containing information letters that included ten patent copies of orbital propulsion etc. to Mr. Lambert of the Professional Engineers Association, with the hope that he can assist me in the finding of an engineer capable of doing the calculations for the equations of motion for orbital propulsion.

Phoned Donna McDuff, physicist at B.C.I.T. and arranged for an appointment for 10:30 a.m., Friday, October 27, 1989.

Thursday, October 26, 1989

Received a phone call from Mr. Len Juteau (around 9:30 a.m.), Department of Advanced Education, Victoria, B.C., phone 387-5028 (TAP) asking me if I had received my $2,500.00 back from the B.C. Research Corporation. When I replied, not yet, he told me that he had been in touch with B.C. Research and they had told him that they would repay the $2,500.00 to me. Mr. Juteau then said that he would call me back in a week's time to see if I had received my money from them. (Apparently they were worried too).

Received envelope by courier around 10:30 a.m. from B.C. Research Corporation. (Copy of letter in file).

Friday, October 27, 1989

Had a meeting at 10:30 a.m. at B.C.I.T. with Donna McDuff discussing the possibility of B.C.I.T. doing work on orbital propulsion and was advised by her that she would phone me the following Friday morning (November 3, 1989). NOTE: She did not call back.

Phoned Dr. Terry Howard's secretary at B.C. Research requesting a mutual exchange for their letter and my cheque. (Copy of letter enclosed).

The meeting was arranged for 3:00 p.m., Monday, October 30, 1989.

Monday, October 30, 1989

Had a meeting with Dr. Howard at B.C. Research. Received $2,500.00 refund in exchange for their letter which had to be signed by me. I refused any further discussions with him regarding further arrangements to supply the equations of motion for orbital propulsion. I believe the only reason I received my $2,500.00 refund was because pressure was put on the B.C. Research by the Minister of Advanced Education (TAP program).

Received a phone call from Mr. Rob McDuff, husband of Donna McDuff (B.C.I.T.) expressing interest in orbital propulsion. I told him that if he was interested further, he could come to my home and discuss it with me.

Thursday, November 2, 1989

Received a phone call from Len Juteau, Victoria, B.C. wishing to confirm the return of my $2,500.00 from B.C. Research. I told him that I received the money.

Monday, November 27, 1989

Had a meeting at 10:30 a.m. with David Dodge and Eric Dunn, astronomers with Southam Observatory in Vancouver. Discussed with them, the possibility of providing the equations of motion for orbital propulsion. I offered a $1,000.00 donation towards their new telescope if they could provide these equations, but they replied that they did not have the expertise to provide these equations.

Wednesday, November 29, 1989

Had a meeting with Richard Mardon, Marketing Director for Finale in Vancouver, discussing the possibility of them producing computer graphics for an orbital propulsion powered spaceship. I was advised by him that he would phone me on Monday, December 4, 1989. He never called back.

Thursday, November 30, 1989

Had a meeting with John Rix, computer graphic artist with Gastown Post and Transfer in Vancouver. Discussed with him the cost of producing computer graphics for orbital propulsion spaceship. He gave me some cost estimates and additional advice. It was too expensive for me. I couldn't afford it.

In the early part of July, 1990 I received a phone call from a man by the name of Mr. Bull who said that he had been talking to Al Knight, an acquaintance of mine who had told him about my orbital propulsion test machine. Mr. Bull asked permission to come over to my home along with a friend of his by the name of Wayne McLeod, to view the prototype model that I had used to test orbital space propulsion. I invited them over and after having coffee, I took them out to the garage to view my test machine. Mr. Bull obviously didn't understand it, but Wayne McLeod, who said he is a draftsman, seemed to grasp most of it. They left and I thought that this was the end of it.

Shortly after, I was surprised to receive a letter from Wayne McLeod, which is as follows:

July 30, 1990

Dear Mr. Knap:

I admit to having been puzzled when I saw by your device on Saturday, but I understand it now. My problem was in resolving how increasing the length could increase the centrifugal force. This would normally slow down the arm, but if it is not allowed to slow down and increase in linear velocity of the weight increases the centrifugal force. Without knowing the math this must have been an act of pure intuition on your part.

Here's a math example I've worked out. I use a radial arm r = 1.5 ft., weight = 0.5 lb. revolving at 5 Rev/Sec. The circumference is thus π x diameter = $^{22}/_{7}$x 3 = 9.4 ft. with 5 revolutions the distance travelled by the weight in 1 second is 9.4 x 5 = 47 ft. So the velocity is V = 47 ft./Sec. Weight must be translated into mass, M = $^{0.5}/_{32}$ = 0.015 (mass) slug. The centrifugal force is given by the formula:

$$F = \frac{MV^2}{r} = \frac{0.015 \times 47^2}{1.5} = 22.09 \text{ lb.}$$

Let's now suppose r increases on the upper portion of the curve to r = 1.6 ft. Now the figure become:

C = $^{22}/_{7}$ x 3.2 = 10 ft. Distance travelled in 1 second = 10 x 5 = 50 ft. The velocity is then V = 50ft/Sec.

$$F = \frac{MV^2}{r} = \frac{0.015 \times 50^2}{1.6} = 23.44 \text{ lb.}$$

Since the bottom weight is travelling at the original arm length, it's centrifugal force is still 22.09 lb. The net difference up is 23.44 - 22.09 = 1.35 lb. + the rocket thrust.

Congratulations!

Yours Truly,
Wayne McLeod

This is the first set of equations that I have ever received from anyone. These equations are from a man who is a draftsman, not an engineer or scientist. This is a man who did this to help me out, for no fee. I will always be grateful to Wayne McLeod.

STILL TRYING

One day in July when I was thumbing through the Yellow Pages in the telephone book, my eye caught the section regarding tutoring. After some thought about the tutoring, I wondered if they would be able to find someone who could help me solve the problems I had with the equations of motion for orbital propulsion. So accordingly I phoned all those in the business and requested a tutor that was skilled in mathematics and calculus. After conversing with the lady of the Teachers Tutoring Society, she gave me the names and phone numbers of three tutors, which was all that she had available that had the specific skills I required. Incidentally, this was the only tutoring service that had the requirements I needed.

The first tutor that I called for an interview was a man by the name Fariborz Fararooni, an Iranian civil engineer, who is a recent immigrant to Canada. This man has degrees from University of Illinois B.S. 1975 and M.S. from Bradley University, Peoria, Illinois 1976 in the United States. I had a meeting with Fariborz at my home. He is a dark-haired, medium height and built man who is articulate, friendly and has an accent. I showed him my material pertaining to orbital space propulsion. Before the meeting was over I had showed him my test machine for orbital propulsion, which we both looked over very carefully. I must admit that he was intelligent enough to ask me pertinent questions about it's operation, including all the test results and measurements that I had done many years ago on this same machine. He took this information home with him and within a few days he phoned me, telling me that he was ready to submit the calculations for equation of motion for orbital space propulsion.

I told him, on our first meeting, that I did not understand higher mathematics, having only gone to the eighth grade in school, so he very carefully explained to me what his calculations meant. The most important thing he told me was that the thrust ratio could be increased by increasing the velocity of the rotating nozzles and also by increasing the mass of the rocket nozzles. These two factors would enable me to obtain a maximum of a 15 to 1 ratio, whereas in my previous testing, I was only able to get a 2 to 1 ratio. Fariborz, who prefers to be called Fred, which I will do from now on, whose set of calculations was extremely important to me if they were correct. It would mean a huge increase in the efficiency of the orbital space propulsion engines, by the reason of using a lot less chemical fuel which is entirely expendable, so that I would get my forward thrust to a much greater degree from the unbalanced centrifugal forces generated by these engines.

I asked Fred for his bill for doing this work and he gave me a figure of $130.00. I said to him that I had been trying to get this information for years and no one could or would supply it to me for love or money. I also said, that here you give me those calculations for a mere $130.00, so I told Bea to write him a cheque for $200.00. Could you imagine, my amazement, joy and gratitude to this man, who was the only qualified person who understood what I had done and the only one who wanted to understand what I had done. Fred was pleased that I appreciated what he had done.

Before the meeting was over we decided to meet on a regular basis in the future, as he knew that these calculations had to be checked over by other qualified experts and he also understood that I could not put these calculations in my book without being confirmed by other experts in the field.

After Fred left I called the other two tutors on my list and made separate appointments to see them. I pretty well used the same procedure that I used with Fred, but I made a point of **not** showing them Fred's calculations as I wanted independent collaboration on this. After some delay, I finally heard from the other two tutors, one who let me know by letter and one in a phone call indirectly to me, that they were unable to do these calculations. So what else is new?

My next move was to phone the Professional Engineers Society in Vancouver to see if I could hire an engineer to check Fred's calculations. After a number of referrals I was told to phone Klohn Leonoff, a large civil engineering company situated in Richmond, B.C. I phoned this company and arranged a meeting with Peter C. Lightall, a professional engineer, who is also manager of the Pacific Region for this firm. In the week of August 16 - 19, 1991, I told Mr. Lightall that I wanted to have the equations of motion for orbital propulsion done. I handed over, to him, pertinent data including Fred's calculations. Peter Lightall turned this job over to a young mechanical engineer by the name of Ty Russell, who worked for his firm. I met Ty Russell on August 20th and I must admit that I was impressed with his intelligence. I told him that I had hired another engineer to do these equations and that I needed another set of calculations from him for comparison. On Wednesday morning August 21st, I went to pick up the equations that Mr. Russell said were ready. What he had told me the day before was drastically changed. From his exuberant very high thrust ratio yesterday, he went to a low thrust ratio today, even lower than my test results. He also said that he didn't believe my test results and started talking about the use of magnets instead of rocket thrust. He was giving me the same old garbage that others had, magnets mean nothing in this case. This upset me so I wanted to make arrangements with him to meet with Fred to compare these equations. He refused to do so. This gave me grave doubts about his engineering ability.

On my return home, I phoned Fred and told him of the several points of the equations that Mr. Russell had mentioned to me. Fred immediately disputed these points. Fred said he could come over to my home Thursday morning, August 22nd.

After I got off the phone I noticed, that when I was looking over Ty Russell's set of equations, that the paper he wrote these equations on was not even signed by him, nor was it written on his firm's letterhead paper. So I decided to phone the Professional Engineers Society to find out whether or not Ty Russell was a registered engineer. He wasn't. Then I phoned Klohn Leonoff and made inquiries regarding Ty from one of the managers. It seemed to upset him that I was asking questions about Ty. He told me that Ty was just a student working there for the summer holidays. A short time later Ty Russell phoned me, angry that I had not called him first. So I told him that I had just found out that he was only a student and not a registered engineer and that I was not going to pay for his equations unless he put them on Klohn Leonoff's letterhead paper and signed his name to it, stating that he was a professional engineer. These equations that I needed for orbital propulsion would only be useful to me if it was signed by a qualified engineer of physicist. Today, August 22nd, Fred came over to my home around 10:00 a.m. and looked over Ty Russell's equation paper. After checking over Ty's equations, Fred told me that they were not correct and were amateurish to boot. I never received a bill from Ty Russell or his firm.

My next try to obtain confirmation of Fred's calculations was to phone the Association of Professional Engineers & Geoscientists of B.C. where I was told to try a company by the name of Dynamic Signal Analysis Corp., an engineering firm in Burnaby, B.C. After phoning for an appointment, I met with several engineers at this office. After discussing what I required, they informed me that what I wanted was beyond their expertise. I asked them if they could help me find the right organization to check over these calculations. A short time later I received a letter from them advising me to write to the Spar Aerospace Limited located at Weston, Ontario and the Bristol Aerospace Ltd. of Winnipeg, Manitoba. I decided to write to both of these aerospace companies. I sent two identical large envelopes containing all the pertinent information regarding orbital propulsion including a set of Fred's calculations. A short time later I received a reply from Spar Aerospace. The following is their reply:

September 25, 1991

Dear Sir:

Thank you for your letter of September 18, 1991 concerning the "orbital propulsion apparatus". Allow me to advise that our field of work does not involve propulsion systems. Further, with regret, I am unable to refer you to a company interested in this specific field.

Please find enclosed with this letter the patent documentation you submitted.

Yours very truly,

A.F.H. Jones
Director
Manufacturing

This letter is from a company that supplied the robot arm on all the United States space shuttles and manufactures satellites. This company is also Canada's largest space company, supported by the Canadian Government and NASA. Spar Aerospace also returned my original letter with my other documents. It seems that they do not want a record of me on file.

MORE POLITICAL GARBAGE

During our Provincial Election Campaign, I received a phone call from an official of Joan Sawicki's campaign headquarters in Burnaby, asking me for voting support for their New Democratic Party (NDP). I immediately replied by asking them for their support. The caller was somewhat puzzled by my reply, so I told him to have Joan Sawicki phone me. The next day Joan phoned me and I gave her a general outline of my problems with Svend Robinson. She appeared concerned and referred the rest of the conversation to David Pepper, who in turn, gave me the telephone number of Audrey McLaughlin, leader of the New Democratic Party (NDP) of Canada, in Ottawa.

I then telephoned Audrey McLaughlin in Ottawa and we discussed my complaint against Svend Robinson. I asked for her fax number so that I could forward letters and other material regarding this matter. Enclosed was the following letter that I sent to Audrey McLaughlin:

September 30, 1991

Dear Madam:

As per our telephone conversation on September 30, 1991, I formally request that you speak to Svend Robinson, my M.P. in Burnaby, B.C. as to why he would not represent me in the matter of orbital space propulsion and my quest for official recognition from the Canadian Government.

I believe that there exists a conspiracy between the Canadian and United States Governments to deny me recognition for my discovery.

Thanking you very much for your time.

I remain,

Yours Faithfully,
George Knap

Fax Number 995-6803
Area Code 613

CONCLUSION

In regards to orbital space propulsion, for over twenty years I have spent considerable time and money, as well as going into every conceivable direction that I could think of in trying to get my discovery off the ground. So I have come to the conclusion that:

a) It could be due to the ignorance and the uncaring attitude on the part of scientists and engineers employed by industry and government and the teaching profession. But I find this hard to believe.
b) There is a long term organized effort to stop me from publicizing my discovery and invention on inter-government levels, possibly by using the Official Secrets Act of Canada and/or the Official Secrets Act of the United States. I am inclined to believe the latter.

I find it very hard to believe, that of all the letters I had written to the various officials of the Canadian and American Governments, newspapers and the industries associated with space technology, that there is little, if anything, that mentions orbital propulsion. I also realize that if the Canadian and/or American Governments have been working on my discovery of a new law in physics as well as orbital space propulsion that vast improvements to my discovery and patent must have been made by now. Yet, when I hired the University of British Columbia to make a citation search, claims citation, USPA, nothing was found in the checking of my patent number on orbital propulsion for citations in the world patent index. In addition, the series of Scisearch Databases were checked to see if any scientific or technical literature cited this patent number. The results were **zero**.

This in itself is strange. All these years and not **one** reference to my patent re-orbital propulsion from the United States patent office. As I have applied for a number of patents myself, I have always received several references regarding the subject patent. It is extremely unusual to have no references at all.

All I have ever wanted was recognition for discovering a new law in physics as well as my invention of orbital space propulsion.

This book is my last ditch effort to gain this recognition before I die.

So I ask myself, is this the end or just the beginning?

REFERENCE MATERIAL

UNCLASSIFIED
OPERATIONS MEMORANDUM

TO: Department of State
FROM: Amconsul Vancouver
SUBJECT: New Rocket Vehicle Invention
DATE: August 7,1968

Attached are two drawings of a new rocket vehicle invention submitted by George Knap, a Canadian citizen residing at 4408 Wildwood Crescent, South Burnaby, British Columbia, for transmission to N.A.S.A.

Also enclosed are thermofax copies of a letter to Mr. Knap from the British Columbia Research Council relating to his invention.

Mr. Knap is a local inventor of some note and holds a number of patents on the things he has invented. He has not attempted to patent this rocket vehicle, which he considers a new method of space propulsion, because the mathematics of his invention have not been completed as to determine the practicality of his idea. Mr. Knap has asked that these documents and the letter from the British Columbia Research Council be sent to N.A.S.A. to determine the feasibility of his new invention with the proviso that N.A.S.A. give his invention every protection and that they furnish him with all mathematical calculations at a later date whether his idea is good or bad so that he can work out the "bugs" in this new rocket vehicle invention.

Sincerely yours,

Aaron R. Coleman
United States Consul

Encls.

NATIONAL AERONAUTICS AND SPACE ADMINISTRATION
WASHINGTON, D.C. 20546

IN REPLY REFER TO: I/ITA

AUG 19 1968

Mr. George Knap
4408 Wildwood Crescent
South Burnaby, British Columbia
Canada

Dear Mr. Knap:

The American Consulate in Vancouver forwarded us your drawings of a rocket vehicle invention and copies of a letter to you from the British Columbia Research Council relating to your invention.

On the basis of your design sketch alone, the Inventions and Contributions Board of NASA would not be able to make a technical evaluation of your invention. We are therefore returning your drawings and suggest that if you should wish to resubmit your design that it be accompanied by adequate supporting detail and specifications.

Sincerely yours,

John P. Riley
John P. Riley
International Technology Affairs Division
Office of International Affairs

DEPARTMENT ...
AIR FORCE ROCKET PROP...
EDWARDS AIR FORCE B...

REPLY TO ATTN OF: CA

9 NOV 1978

SUBJECT: Evaluation of Patent - "Orbital Propulsion Apparatus"

TO: Acme Research and Development Co., Ltd
ATTN: Mr. George Knap, Manager
4408 Wildwood Crescent
South Burnaby, B.C. Canada

Dear Mr. Knap

1. Your patent application has been reviewed and evaluated by technical personnel of this Laboratory. It appears that your proposed apparatus attempts to develop additional propulsive force in violation of Newton's laws of motion. The laws imply that the sum of the momentum vectors of all masses which remain with the vehicle must always be equal and opposite to the momentum vector of all masses which leave the vehicle. The only mass that leaves your vehicle is the propellant expelled by the rocket motors. The rotating machinery does not increase the momentum of the expelled propellant, but rather serves only to vary the relative position and velocity of masses which remain attached to the vehicle. Consequently, no improvement in range or engine performance is anticipated through the use of your apparatus.

2. We appreciate the interest and time that you at Acme Research and Development Co., Ltd, have displayed in submitting this proposal. One copy of your application is being retained in our files for future reference. Your original copy is being returned to your company.

Sincerely

Richard R. Weiss

RICHARD R. WEISS
Chief Scientist

1 Atch
Original

NASA

National Aeronautics and
Space Administration

Washington, D.C.
20546

Reply to Attn of: RP-4

DEC 4 1981

Mr. George Knap
4408 Wildwood Crescent
South Burnaby, B.C.,
Canada V5G 2M4

Dear Mr. Knap:

We have a continual interest in future space power and propulsion systems. We assess high potential advanced concepts and demonstrate the technical feasibility of systems that offer great promise.

We examined a copy of your patent for an Orbital Propulsion Apparatus which was forwarded by Senator John Glenn. Your concept does not appear to offer any advantage over the ideas we are presently investigating. We are not studying similar concepts. Under the circumstances, we will not be interested in pursuing your idea in the future.

We suggest that you consider publicizing your idea by presenting a paper describing it to an appropriate technical symposium.

C. Robert Nysmith

C. Robert Nysmith
Assistant Associate Administrator
for Management Support

ACME RESEARCH AND DEVELOPMENT CO. LTD.

4408 WILDWOOD CRESCENT SOUTH BURNABY, B.C., CANADA V5G 2M4

TELEPHONE: HEmlock 3-7415

December 28, 1981

Reply to Attn. of RP-4

National Aeronautics and
Space Administration
Washington, D.C.
U. S. A. 20546

Attention: C. Robert Nysmith
Assistant Associate Administrator
for Management Support

Dear Mr. Nysmith:

In reply to your letter of December 4, 1981, I hereby request the Analytical Data pertaining to my invention titled "Orbital Propulsion" U. S. A. Patent Number 4,087,064 i.e., Rocket Thrust to Engine Output Ratios, combined to produce propulsion, which I assumed you have done.

In order to prepare a paper describing this new principal, I need and request the collaborating scientific evidence by an organization of your stature.

This simple prototype machine was built by myself to test this new principal. The results were obtained by me with the assistance of a group of scientists from the Simon Frazer University of British Columbia. As you know ideas and opinions do not count for much in the scientific community.

Yours truly

George Knap

GK/et

NASA
National Aeronautics and
Space Administration
Washington, D.C.
20546

Reply to Attn of: R

Mr. George Knap
4408 Wildwood Crescent
South Burnaby, B.C.
Canada V5G 2M4

Dear Mr. Knap:

I apologize for the delay in answering your letter of December 28, 1981. In your letter you asked for analytical data pertaining to your invention titled "Orbital Propulsion Apparatus", U.S. Patent Number 4,087,064.

In your original letter requesting a review of your invention you did not forward any analytical data related to it. We do not have such data. We only examined that which was available, namely the patent and determined that your concept that you described therein did not appear to offer any advantage over the ideas we were investigating. In a time of severe budget constraint we must consider those concepts we have enough information to determine their potential,and perhaps, their feasibility,given sufficient funding.

C. Robert Nysmith
C. Robert Nysmith
Assistant Associate Administrator for Management
Office of Aeronautics and Space Technology

Office of the
Minister of State

Science and Technology
Canada

Cabinet du
Ministre d'État

Sciences et Technologie
Canada

Ottawa, Canada
K1A 1A1

March 4, 1985

Mr. George Knap
George Knap Enterprises
4408 Wildwood Crescent
Burnaby, British Columbia
V5G 2M4

Dear Mr. Knap:

On behalf of the Honourable Tom Siddon, I would like to thank you for your letter of February 27, 1985 requesting a meeting with the Minister to discuss new technology in transportation, specifically the Knap patent on orbital propulsion apparatus.

Please be assured that your letter will be brought to the attention of the Minister at the earliest opportunity.

Yours sincerely,

DACameron

Deborah Cameron
Executive Secretary
British Columbia Minister's Office

Canada

United States Patent [19]

Knap

[11] **4,087,064**

[45] **May 2, 1978**

[54] **ORBITAL PROPULSION APPARATUS**

[76] Inventor: George Knap, 4408 Wildwood Crescent, Burnaby, British Columbia, Canada, V5G 2M4

[21] Appl. No.: 820,146

[22] Filed: Jul. 29, 1977

Related U.S. Application Data

[63] Continuation-in-part of Ser. No. 727,036, Sep. 27, 1976, abandoned, which is a continuation-in-part of Ser. No. 623,388, Oct. 17, 1975, abandoned, which is a continuation of Ser. No. 452,078, Mar. 18, 1974, abandoned.

[51] Int. Cl.² B64G 1/20

[52] U.S. Cl. .... 244/172; 74/84 S; 244/62

[58] Field of Search .... 244/158, 167, 169, 172, 244/62; 74/84 S

[56] **References Cited**

U.S. PATENT DOCUMENTS

3,423,755 1/1969 Lassen et al. .... 244/158 X

Primary Examiner—Barry L. Kelmachter

Attorney, Agent, or Firm—Carver and Company

[57] **ABSTRACT**

Orbital propulsion apparatus which has a rotor drivable by an electric motor and upon which is mounted a plurality of mass members for movement radially of the axis of the rotor, each mass member having a rocket motor mounted thereon which, when fired, moves its associated mass member outwards against a compression spring mounted on the rotor so that at a predetermined angular velocity of rotor the mass members rotate in a common circular orbit, the rockets being controllable so that they can be fired through an arc of predetermined angularity so as to move each mass member radially outwards of its circular orbit to increase the centrifugal force of each mass member across the arc while the electric motor operates to maintain constant angular velocity of the rotor.

1 Claim, 8 Drawing Figures

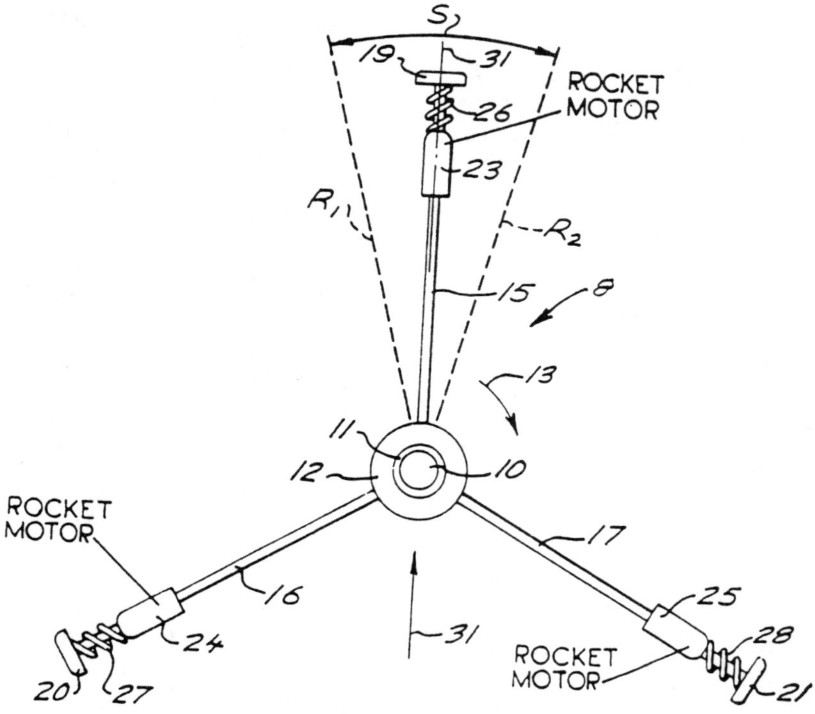

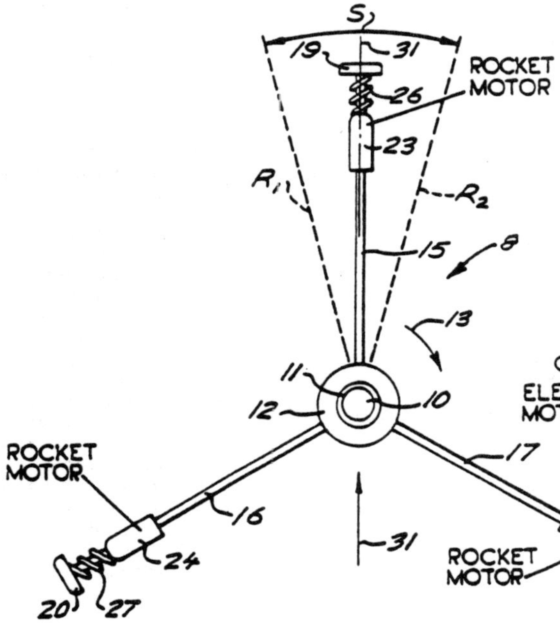

—Fig 1—

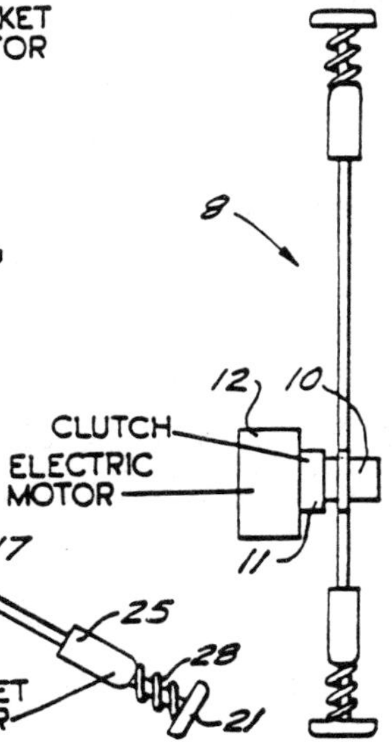

—Fig 2—

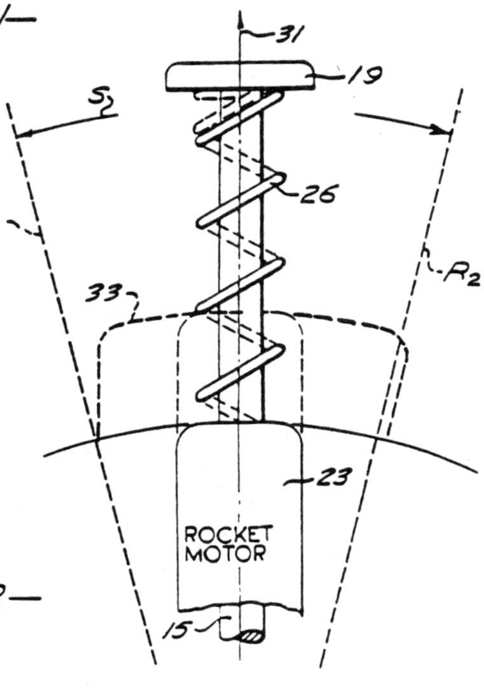

—Fig 3—

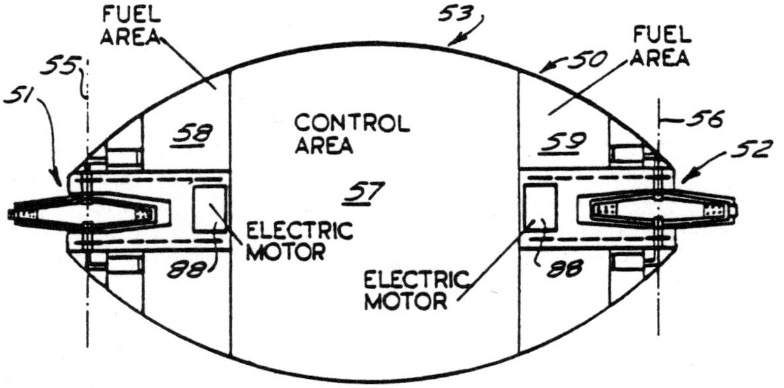
FUEL
AREA
53
55
51
50
FUEL
AREA
56
58
CONTROL
AREA
59
52
57
ELECTRIC
MOTOR
88
ELECTRIC
MOTOR
88

Fig. 4

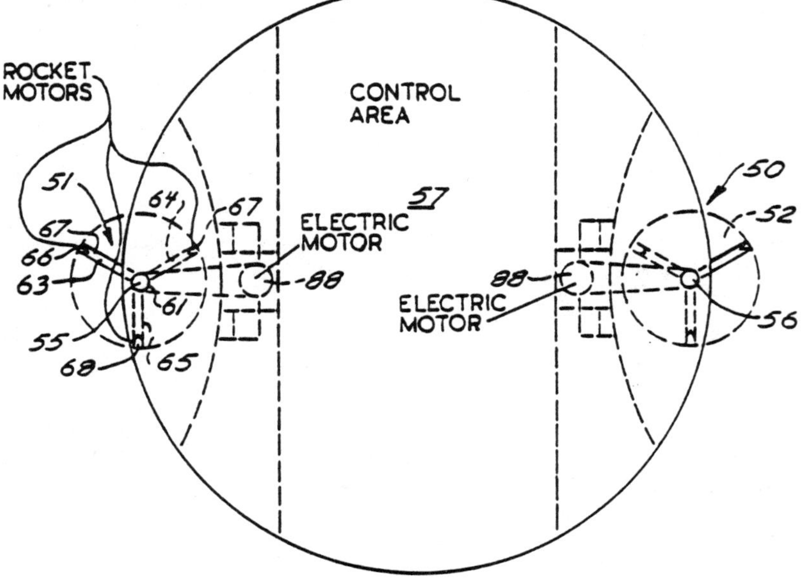
ROCKET
MOTORS
CONTROL
AREA
50
57
51
64
67
ELECTRIC
MOTOR
52
67
66
63
88
88
ELECTRIC
MOTOR
61
56
55
68
65

Fig. 5

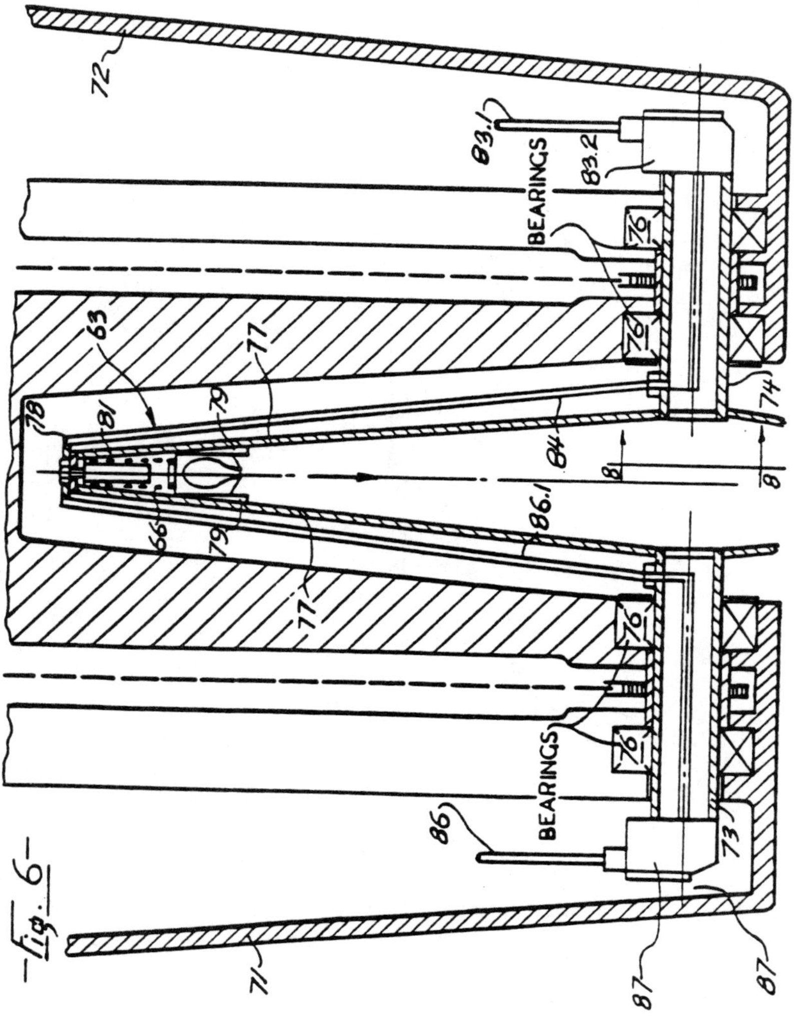
72
83.1
83.2
BEARINGS
76
76
74
63
77
78
81
79
84
8
8
86.1
66
79
77
76
76
BEARINGS
86
73
87
87
71

Fig. 6

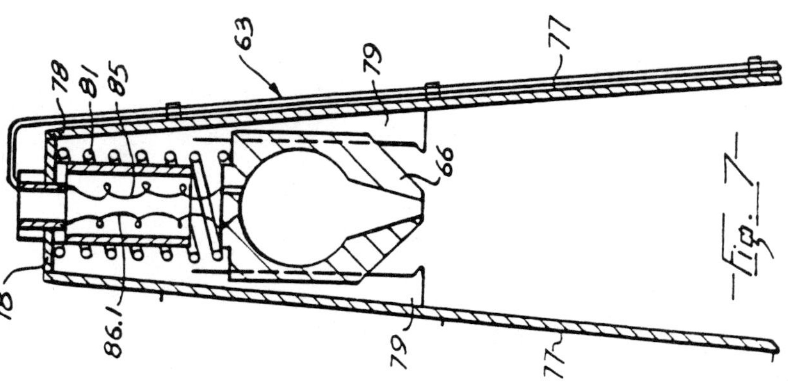
63
77
79
78
81
85
66
78
86.1
79
77

Fig. 7

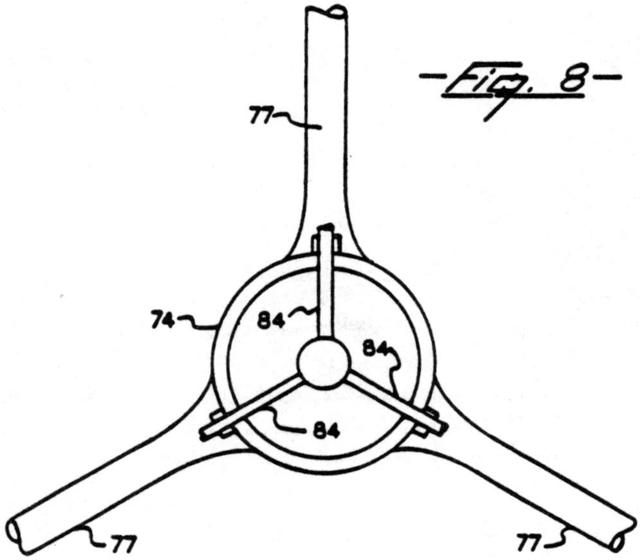

-Fig. 8-

ORBITAL PROPULSION APPARATUS

CROSS REFERENCES TO OTHER APPLICATIONS

This is a Continuation-in-Part of my application Ser. No. 727,036 filed Sept. 27, 1976 entitled ORBITAL PROPULSION APPARATUS, now abandoned which is a continuation-in-part of Ser. No. 623,388, Oct. 17, 1975, now abandoned, which is a continuation of Ser. No. 452,078, Mar. 18, 1974, now abandoned.

BACKGROUND OF THE INVENTION

Field of the Invention

This invention relates to propulsion apparatus particularly directed to but not limited to apparatus for propelling vehicles in airless regions.

Propulsion apparatus for travel in airless regions, at the present date, is limited to chemically fueled reaction engines, namely rockets. Although this type of engine is efficient, weight and volume of rocket fuel necessarily limits the range and payload of vehicles propelled in this manner.

Unless use can be made of energy developed by atomic reaction to afford a propulsive effect, space craft will only be able to travel a relatively short distance away from the earth. To date no reaction engine has been developed utilizing the energy released by atomic reaction. Engines, however, have been developed where an energy released by atomic reaction in the form of heat can be utilized to provide electrical energy.

SUMMARY OF THE INVENTION

The present invention provides propulsive apparatus wherein electrical energy developed through atomic or similar reactions can be utilized to effectively aid in propulsion of vehicles in airless regions.

In the apparatus of the present invention orbital diameters or orbiting masses in balance are selectively shifted to provide increased centrifugal force in a selected direction and tendency of increased orbital diameter of the orbiting masses to result in a decrease in angular velocity is overcome by force applied by an electric motor the energy of which can be developed through atomic reaction. As atomic fuel has little mass relative to the energy available the vehicle propelled by the present invention can have a range much in excess of vehicles propelled solely by rocket fuels.

A detailed description following related to the drawings gives exemplification of apparatus according to the invention which, however, is capable of expression in means other than those particularly described and illustrated.

DESCRIPTION OF THE DRAWINGS

FIG. 1 is a front elevation, mainly diagrammatic, of one embodiment of the propulsion apparatus in accordance with the invention,

FIG. 2 is a side view thereof,

FIG. 3 is an enlarged front elevation detailing a portion of the apparatus shown in FIG. 1,

FIGS. 4 and 5 are side and plan sectional views, mainly diagrammatic, showing another embodiment of the invention,

FIG. 6 is a sectional side view, much enlarged, of a portion of the apparatus shown in FIGS. 4 and 5,

FIG. 7 is a portion of FIG. 6, much enlarged,

FIG. 8 is a section taken on Line 8—8 of FIG. 6.

DETAILED DESCRIPTION

Referring to the drawings and particularly to FIGS. 1 and 2 one embodiment 8 propulsion apparatus in accordance with the invention has a rotor, generally 9, having a hub 10 suitably mounted for rotation on the vehicle, not shown, to be propelled, the hub being connected through an overriding clutch 11 to an electric motor 12. The clutch permits the motor to drive the hub uni-directionally, shown by the arrow 13.

Electric power for driving the electric motor is provided, preferably, by a nuclear reactor and attendent generating components which, for example, can be the type now in use for driving atomic powered submarines and the like. The electrical generating equipment is, of course, carried in the craft to be propelled.

Three rods 15, 16 and 17 extend radially from the hub and terminate in stops 19, 20 and 21. The rods which are of equal length are symmetrically disposed about the hub.

Rocket motors and mass members 23, 24 and 25 are mounted on the rods 15, 16 and 17 respectively for slideable movement longitudinally of the rods. Compression springs 26, 27 and 28 are mounted over the rods between the mass members thereon and the stops. The rocket motors which carry their own fuel are arranged so that when fired the rockets exert a thrust radially outwards. The rockets are all of the same weight and the springs are identical. The rockets are of a type which can be fired intermittently, control of period and location of firing being obtained by suitable controls within the vehicle.

OPERATION

In operation, the rotor is driven by operation of the electric motor until the rotor reaches a predetermined angular velocity. The rockets and mass members, through centrifugal action are moved outwards against the springs compressing the latter slightly to rotate in a common circular orbit. As the weight of the rocket and mass members and strength of the springs are the same, centrifugal forces balance each other so that the apparatus is in a state of equilibrium.

In order to achieve movement of the propulsion apparatus and with it the vehicle in a direction indicated by the arrow 31 each of the rockets is fired, in turn, as it passes through an arc S of its orbit the extent of which is defined by lines of radius R1 and R2 which subtend an arc of about 20° bisected by the line 31 representing the direction of proposed travel. Thrust generated by each rocket, as it is fired, moves each associated mass member from its normal position, as shown in solid outline in FIG. 3, radially outwardly along its rod to an extended position shown in broken outline in FIG. 3, thus increasing its orbital radius, the extended orbit being defined by the broken line 33. Normally, with inertia constant, increase in orbital radius of each rocket and mass member, will result in a decrease of angular velocity, however, constant angular velocity is maintained through operation of the electric motor during the period in which the firing rocket passes through the orbital arc S. When each rocket ceases firing as it leaves arc S the spring will again return the rocket and mass member to its normal orbit.

Propulsive effect, it is seen, is gained, not only from the thrust of each rocket but also from increased centrifugal force of the rocket and mass members due to the increased orbital speed of each rocket as it passes through the arc S. The increased centrifugal force along line 31 is equal to the energy output of the electric motor during the passage of each rocket and a mass member through the arc S.

Operational effectiveness is largely dependent upon synchronous relationship of its many parts, however, for optimum effectiveness, angular velocity, angle of orbital arc S, natural frequency of the springs and periodic operation of the electric motor should be so related so as to take advantage of cumulative effect of cooperating parts operating at resonant frequencies.

Another embodiment 50 of the invention is shown in FIGS. 4 through 8.

Referring to FIGS. 4 and 5 a pair of orbital propulsion motors 51 and 52 are shown mounted in a craft 53 on parallel axes 55 and 56. The craft as shown has a control area 57 and fuel areas 58 and 59 at ends of the craft.

The motor 51 has a hub 61 to which is secured three symmetrically arranged radial arms 63, 64 and 65. Rocket engines 66, 67 and 68 which are identical in structure and weight are mounted on the arms 63, 64 and 65, respectively.

FIGS. 6 and 7 show, in detail, one of the arms 63 together with rocket engine 66 of the propulsion motor 51 mounted for rotation between a pair of spaced brace members 71 and 72 extending from the hull of the craft. The hub of the motor 51 comprises a pair of hollow axles 73 and 74 mounted for rotation in bearings 76 secured to the brace members 71 and 72. The arm 63 is formed of a pair of spaced outwardly converging struts 77, one of the struts having a rigid connection to the axle 73 and the other strut having a rigid connection to the axle 74 and are connected at their outer ends by a bridging piece 78.

A pair of parallel rails 79 — 79 are secured to the struts near the outer ends thereof and slideably support the rocket engine for radial movement. A compression spring 81 fits between the rocket engine 66 and the bridging piece 78. Rocket fuel is delivered from the fuel tanks in the craft to the rocket engine 66 through a conduit 83.1 which is connected to a swivel 83.2 at one end of the axle 74 thence through a conduit 84 which extends through the axle and outwardly along the arm 63 and thence by flexible conduits 85 to the rocket engine. Electrical power to the rocket engine 66 is fed through lines 86 which are connected to a rotary connector 87 on the axle 73 thence through lines 86.1 extending through the axle 73 and along the arm 63 to the rocket engine. The arms 64 and 65 and the rocket engines thereon together with fuel and electrical lines are as described with reference to the arm 63.

Blast from the rocket engines it is seen passes between the axles 73 and 74 while thrust of the rocket engines is radially outward against their respective compression springs.

The axles of the motors 51 and 52 are connected for driving engagement to electric motor 88 mounted in the craft. The motor 52 is identical to motor 51 and, consequently, it is not considered a detailed description thereof is required. Period of operation of the electric motors, delivery of fuel to the rockets and fuel ignition can be controlled by controls within the craft such that the rocket engines can be fired through any desired arc as the rotors rotate.

Operation of embodiment 50 is the same as described with reference to embodiment 8. In embodiment 50, however, provision of the motors provides balance control of the craft for operation thereof in any direction about an axis parallel to the axes of the rotors. Control of attitude of the craft so as to alter its direction of travel can be effected by operation of control rockets fed by the same fuel as used by the rocket motors 51 and 52. These control rockets are not shown as such devices are of common knowledge.

I claim:

1. Orbital propulsion apparatus comprising:

(a) a rotor having a rotational axis,

(b) a plurality of arms radiating from the rotor,

(c) a mass member mounted on each arm for rotation therewith and freely movable longitudinally of each arm,

(d) yieldable resilient means reacting against the outer end of each arm and each mass member normally urging the mass members radially inwards for resiliently restraining radial outward movement of the mass members so that the mass members normally rotate in a common circular orbit when the rotor rotates at a pre-determined velocity,

(e) a reaction motor mounted on each mass member for urging each mass member to move radially outwards against the action of its associated resilient means, each said motor being controllable for effecting its operation through an arc of predetermined angularity,

(f) motor means drivingly connected to the rotor and synchronized with operation of the reaction motors to maintain rotor speed constant so that the mass members return, under the action of the resilient means, to the common circular orbit when the reaction motors cease operation.

* * * * *

Office of the
Minister of State

Science and Technology
Canada

Cabinet du
Ministre d'État

Sciences et Technologie
Canada

Ottawa, Canada
K1A 1A1

March 4, 1985

Mr. George Knap
George Knap Enterprises
4408 Wildwood Crescent
Burnaby, British Columbia
V5G 2M4

Dear Mr. Knap:

On behalf of the Honourable Tom Siddon, I would like to thank you for your letter of February 27, 1985 requesting a meeting with the Minister to discuss new technology in transportation, specifically the Knap patent on orbital propulsion apparatus.

Please be assured that your letter will be brought to the attention of the Minister at the earliest opportunity.

Yours sincerely,

D. Cameron

Deborah Cameron
Executive Secretary
British Columbia Minister's Office

Canada

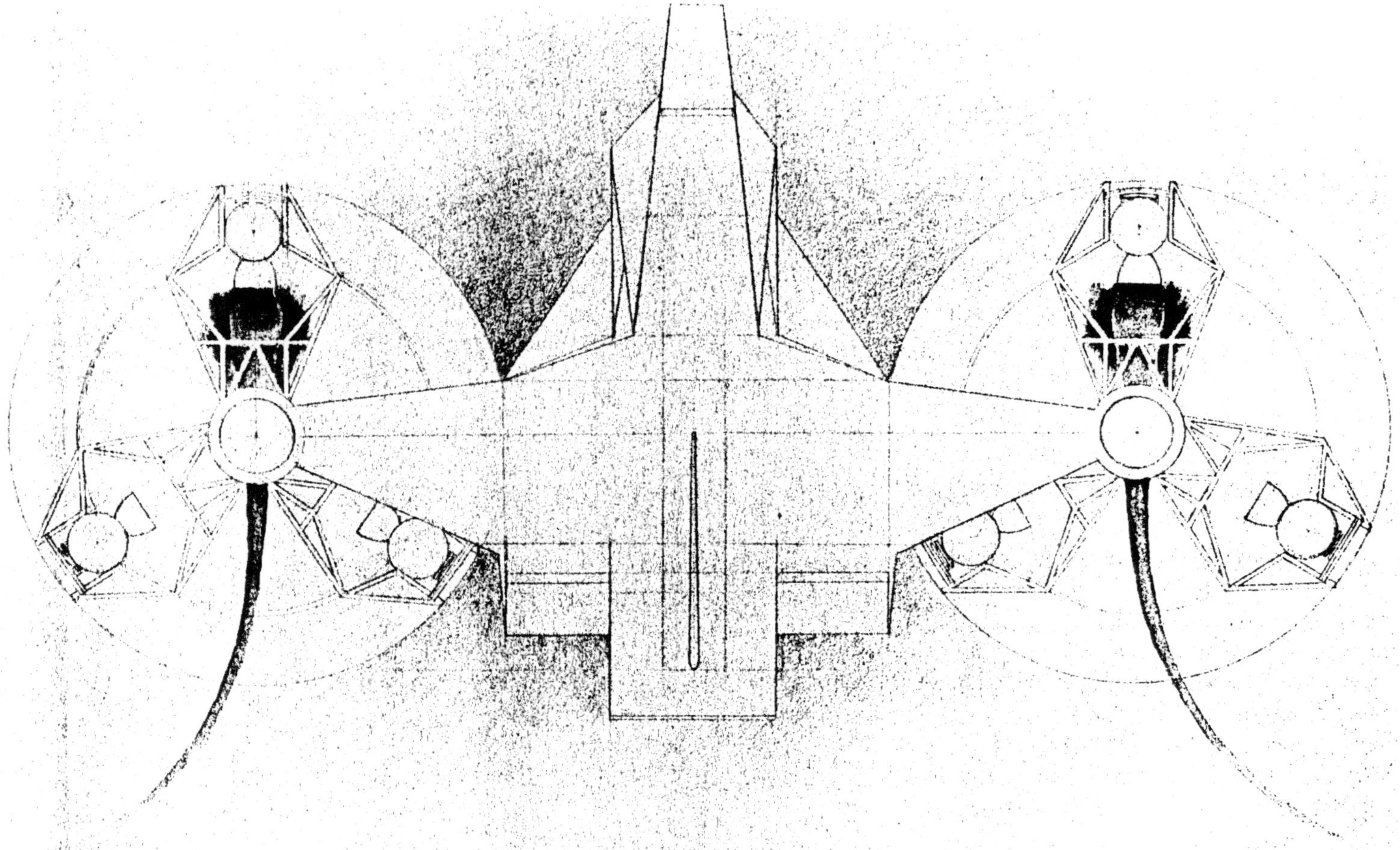

INVENTOR GEORGE KNAP ORBITAL PROPULSION SPACECRAFT ILLUSTRATION BY 9/6/85

GEORGE KNAP ENTERPRISES

UNIQUE DESIGNS, CONCEPTS AND INVENTIONS

4408 WILDWOOD CRESCENT, BURNABY, B.C., CANADA V5G 2M4

TELEPHONE (604) 433-7415

Beginning of
Mail to Politicians
September 17, 1986

Dear Sir:

I wish to inform the media of my discovery of new laws in Physics, namely my invention of "Orbital Propulsion" U.S.A. Patent 4087064 of which a copy is enclosed.

My concept for Orbital propulsion began over 18 years ago. The first five years was devoted to study, then I decided to build a working model to test my theory. Assisting me in these tests were scientists from the Universities of B. C. and Simon Fraser, B. C. As these test results were positive, I filed for a U.S.A. patent.

The capability of "Orbital Propulsion" in the future of space exploration rests in two distinct categories:

1) As a propulsion system for a space shuttle the advantages are single stage propulsion, no attachments such as booster rockets and/or auxiliary fuel tanks. There will be aircraft type lift off and landing, very high fuel efficiency and much greater margin of safety than present means.

2) In the future a form of space liner must be used that will be self contained and be capable of very high speeds and be able to conserve its fuel mass and will use atomic and solar energy as its main fuel source. This vehicle should have a range capable of reaching other star systems.

I am sure that my "Orbital Propulsion" systems will meet these requirements.

Through Acme Research and Development Co. Ltd. (my Company), I sent a letter to the President of the National Research Council requesting assistance for my discovery of Orbital Propulsion. The reply from the new President of N.R.C. stated that the President I had written to was no longer with N.R.C. I was firmly refused assistance in this matter, without them seeing the necessary information.

I then sent a Patent Copy of "Orbital Propulsion" together with a letter offering this invention to NASA. I received a reply together with an application form for a funding grant. I soon realized my small research company could not meet NASA's requirements, (financial ability), so I did not pursue this matter with NASA at that time.

When I applied for the U.S.A. Patent on "Orbital Propulsion" it was through Carver and Co. who was my Patent Agent over a long period of time doing a lot of my other work as well.

Carver's principal Agent (now deceased) for some reason seemed to be trying to prevent this Patent from going through, it took so long. (See Continuation-In-Part on Patent Copy). When I threatened legal action, they finally secured this U.S.A. Patent.

In order to publicize my discovery, I mailed my Patent copies to many foreign consulates. I received a telephone call supposedly from the French Government expressing interest. Also the Australian Government in a letter advising me to contact NASA. I also sent a cable to the Soviet Academy of Science asking assistance. I did not receive a reply so I assumed they were not interested. Also since I never received a copy of the cable I sent or a bill for it, I had to assume the message was not sent. Recently I had the same experience with CNCP Telecommunications on a cable I sent to the European Space Agency in Paris.

Through a business friend of long standing I forwarded Orbital Propulsion information to reach a high level in the Canadian Government. I received a reply from General Dextraze advising me to get in touch with the National Research Council, but being previously rebuffed by the N.R.C., I did not approach them. I also approached the Science Minister for assistance in this matter, to no avail.

I then mailed a number of Patent Copies of "Orbital Propulsion" to major American aero space companies to get their interest, to no avail and no replies. In a moment of desperation I sent a letter to U.S. Senator John Glen requesting his assistance in approaching NASA at a higher level. I did receive several replies advising me that they had superior technology than my Orbital Propulsion system (this was before the space shuttle disaster). NASA advised me to publish my findings in a science paper. Having no university degree to do so, (PhD is required), this was not possible.

On learning that Vancouver was to host EXPO 86 and its theme was to be Transportation and Communications, I made telephone inquiries to EXPO officials so that Orbital Propulsion and some of my other inventions in the field of transportation could be presented for showing. I offered this package deal for free, even though I only had a Patent on Orbital Propulsion and not the others. Not being able to afford the costs of the Patents on my other inventions, I would lose my legal rights by publicizing them at EXPO 86. My main concern really was to publicize my "Orbital Propulsion System".

After numerous attempts to arrange meetings with these officials, I gave up for the present. I then approached Imperial Oil Co. Ltd. (one of EXPO 86 sponsors). With their assistance a meeting was arranged with David Youngson, an official for EXPO 86. At the first meeting he expressed interest in my work and said my work must be presented in a more visual way and gave me two weeks time to engage a suitable artist to have this work done, which was no easy task.

In our second meeting, I presented my illustrations. He then told me he would let me know whether my work would be displayed at EXPO 86. As time went on and his deadline was approaching, I phoned Mr. Youngson and asked about displaying my work. He was very evasive, so I reached the conclusion of being strung along and told him so.

Shortly after, while listening to CJOR radio, Jim Pattison, President of EXPO 86 was discussing complaints about local participants being denied access to EXPO. He promised access to himself for those voicing complaints. So I arranged a meeting with Mr. Pattison and showed him my illustrations of Orbital Propulsion and my other futuristic inventions in the field of transportation. He was very receptive and expressed keen interest in my work and said he envied my inventive mind. He also liked that I was born in Vancouver and had worked on the EXPO site years ago.

After my meeting with Mr. Pattison, David Youngson phoned me, not pleased that I had gone over his head, telling me he was interested in the land tanker only, leaving out Orbital Propulsion which I consider my most important work. This ended up with my saying that if he was not accepting Orbital Propulsion, I was not interested in displaying any of my other inventions at EXPO. I wrote to Mr. Youngson making a final appeal expressing concern about his choice of land tanker over Orbital Propulsion. I sent a copy of this letter to Mr. Pattison as well also thanking him for his assistance. I never received a reply from either person.

In the same period I also approached officials at the Canada Pavilion EXPO 86 trying to interest them in displaying Orbital Propulsion but due to these people having prior knowledge of my problem with David Youngston, they refused to look at my work.

I then arranged a meeting with Tom Siddon's personal assistant (former Science Minister) at his Vancouver office in a last ditch effort to overcome this problem. I was assured at the time by his personal assistant that they would back me at Canada Pavilion. She had known in advance of my problems. After waiting and not hearing from them, I telephoned her to find out what was happening. She informed me that I would receive no help from this Government Dept. or from Tom Siddon and that he refused to see me. This ended all hopes of displaying my invention at EXPO 86.

Due to being constantly rebuffed in every direction over a long period of time and having a lot of my mail and overseas cables not being delivered, I have come to some conclusions:

1) That the Canadian and American Governments are working on my invention in secret thereby conspiring to defraud me of my discovery.

2) My test machine has been available all these years for testing, questioning, etc. All I want to know is, if the Orbital Propulsion does not work, why has no-one explained why it doesn't or even questioned me about it, except for the Russian space experts at EXPO 86.

I accuse the Canadian Government of being grossly negligent in not investigating Orbital Propulsion because of vast implications pertaining to National Defense.

Will the reader please note that the Canadian Government officially acknowledges the contents of this submission and does not dispute my claims.

NOTE:
COPIES WERE MAILED TO ALL MEMBERS OF THE CANADIAN HOUSE OF COMMONS AND SENATE, THE UNITED STATES SENATE THE UNITED STATES CONGRESS

Yours truly,

GEORGE KNAP

ALL INQUIRIES WILL BE ANSWERED

DEFENSE ADVANCED RESEARCH PROJECTS AGENCY
1400 WILSON BOULEVARD
ARLINGTON, VA 22209-2308

OCT 20 1986

George Knap Enterprises
4408 Wildwood Crescent
Burnaby, British Columbia
Canada V5G 2M4

Dear Sir:

Thank you for submitting your patent on "Orbital Propulsion Apparatus" to us for review. Our National Aerospace Plane (NASP) program is interested in the technology leading to a transatmospheric vehicle. As such the time spent in orbit will be limited and so aerodynamic and conventional thrusters will be the control and propulsion methods of choice. As you state the invention is directed to and best adapted to flight in "airless regions" and the NASP is planned to operate in low Earth orbit and the Earth's atmosphere.

The NASA activities with deep space exploration would more likely benefit from application of your invention and they are in the technology for Space business. I suggest you contact Dr. Len Harris at NASA, who is the Director of Space for the Office of Aeronautics and Space Technology. His address is Code RS, NASA Headquarters, Washington, D.C. 20546.

Sincerely,

Allen R. Atkins II
Director
Aero-Space Technology Office

Minister of National Defence CANADA Ministre de la Défense nationale

November 19, 1986

Mr. George Knap
George Knap Enterprises
Unique Designs, Concepts and
Inventions
4408 Wildwood Crescent
Burnaby, British Columbia
V5G 2M4

Dear Mr. Knap:

Thank you very much for your letter of September 17, 1986, in which you enclosed a Patent Copy of "Orbital Propulsion".

I appreciated receiving this document, and I am grateful for the time taken to bring it to my attention.

Again, thank you and best wishes.

Sincerely,

Perrin Beatty

Ottawa, Canada K1A 0K2

Minister of State for Science and Technology | Ministre d'État chargé des Sciences et de la Technologie

CANADA

The Honourable | L'honorable
Frank Oberle

DEC 19 1986

Mr. George Knap
George Knap Enterprises
4408 Wildwood Crescent
Burnaby, British Columbia
V5G 2M4

Dear Mr. Knap:

I have noted with interest your letter of September 17, 1986, describing your invention of "Orbital Propulsion" and the U.S. Patent 4087064.

I can assure you that, since Canada is a contracting member country of the Paris Convention of 1883/1967 related to patents, full recognition to you for priority as of the date of disclosure for the claims registered in U.S. Patent 4087064 will be recognized in Canada. I would note, however, that laws of nature or physics cannot be patented and only the claims described in the patent are part of your invention. Further, I can assure you that no federal agency in Canada would work on the application of your invention without appropriate recognition as to the inventor. It is unfortunate, but must be recognized, that very few inventions find commercial application.

I trust these assurances will be helpful to you and I wish you success in your continued design and development activities.

Yours sincerely,

Frank Oberle

cc: Mr. Raymond Skelly, M.P.

It's our year! C'est notre année!
in motion...in touch | en mouvement...au courant
86

Ottawa, Canada K1A 1A1

ORRIN G. HATCH
UTAH

135 RUSSELL SENATE OFFICE BUILDING
TELEPHONE: (202) 224-5251

HATCH HOT LINE 1-800-662-4300
(UTAH TOLL FREE)

United States Senate

WASHINGTON, D.C. 20510

COMMITTEES:
JUDICIARY
LABOR AND HUMAN RESOURCES
SMALL BUSINESS
BUDGET
OFFICE OF TECHNOLOGY ASSESSMENT

February 12, 1987

Mr. George Knap
George Knap Enterprises
4408 Wildwood Crescent
Burnaby B.C. V5G 2M4, CANADA

Dear Mr. Knap:

I appreciate your taking the time to send me your views.

It is important to me to have the benefit of your insights on the important issues facing our nation.

Sincerely,

Orrin Hatch

Orrin G. Hatch
United States Senator

OGH/svd

United States Senate
WASHINGTON, D.C. 20510

February 17, 1987

Mr. George Knap
4408 Wildwood Crescent
Burnaby B.C. 45G 2M4
Burnaby, CANADA

Dear Mr. Knap:

Thank you for sending me information on your orbital propulsion proposal. It is always a pleasure to learn of ideas that could be beneficial to NASA and the space program.

I appreciate your confidence in providing me with this information; however, I believe that it is in your best interest to directly submit your materials to NASA for their analysis. The address is:

The National Aeronautics and Space
Administration
400 Maryland Avenue, SW
Washington, DC 20546

Thank you again for writing, and please feel free to contact me again if I can be of further assistance.

Best regards.

Sincerely,

John Glenn

John Glenn
United States Senator

JG/alm

United States Senate

COMMITTEE ON COMMERCE, SCIENCE,
AND TRANSPORTATION

WASHINGTON, DC 20510-6125

February 17, 1987

Mr. George Knapp
4408 Wildwood Crescent
Burnaby, B.C.
CANADA V5G 2M4

Dear Mr. Knapp:

Thank you for your letter concerning your Orbital Propulsion Patent.

I have asked my staff to look into this matter as part of our oversight of NASA's space research and technology activities in order to determine its feasibility and applicability to NASA programs.

With warmest personal regards, I am

Sincerely,

Fritz Hollings

ERNEST F. HOLLINGS
Chairman

EFH:mks

PHILIP M. CRANE
MEMBER OF CONGRESS
12TH DISTRICT OF ILLINOIS

WAYS AND MEANS
COMMITTEE

SUBCOMMITTEES:
TRADE
SOCIAL SECURITY

REPUBLICAN STUDY COMMITTEE
EXECUTIVE COMMITTEE

Congress of the United States
House of Representatives
Washington, DC 20515

OFFICES:
SUITE 1035
LONGWORTH BUILDING
WASHINGTON, DC 20515
202/225-3711

SUITE 101
1450 SOUTH NEW WILKE ROAD
ARLINGTON HEIGHTS, IL 60005
312/394-0790

56 NORTH WILLIAMS STREET
CRYSTAL LAKE, IL 60014
815/459-3399
312/223-3030

ROBERT C. COLEMAN
ADMINISTRATIVE ASSISTANT

March 9, 1987

Mr. George Knap
George Knap Enterprises
4408 Wildwood Crescent
Burnaby, B.C. Canada V5G 2M4

Dear Mr. Knap:

Thank you for contacting me and describing your invention to me.

You may consider submitting your idea to local venture capitalists. These businessmen are often quite receptive to new ideas since they are in business to fund new companies. You might also try contacting professors of electrical engineering and request their opinion of your invention in the hope of receiving a letter of recommendation.
I would also suggest that you write to Congressman Doug Walgren, who is the Chairman of the Subcommitte on Sience, Research, and Technology. His address is 2319 Rayburn House Office Building, Washington, D.C. 20515.

Sincerely,

Phil Crane

Philip M. Crane, M.C.

PMC/tg

NASA

National Aeronautics and
Space Administration

Washington, D.C.
20546

MAR 12 1987

Reply to Attn of: RP

Mr. George Knap
George Knap Enterprises
4408 Wilwood Crescent
Burnaby, B.C., Canada V5G 2M4

Dear Mr. Knap:

Your letter of September 17, 1986 to Dr. Harris was referred to our office. I apologize for the delay in answering your inquiry, but we wanted to be sure that we could provide you with a meaningful response.

It is evident that you have spent a lot of time and effort in developing your Orbital Propulsion Apparatus concept. It is surpising that you haven't gotten more outside interest, or at least some answers to your questions.

Our assessment of your concept finds no fault with the mechanical design and agrees that indeed forward thrust would be achieved and that acceleration would be imparted to the spacecraft, but the question is: how much? In our judgement, so much inert mass has been added (nuclear power system and complex mechanical devices) that no matter how efficiently the resulting thrust is produced, the overall effect on spacecraft acceleration is probably less than if it were powered by a conventional propulsive device.

A more precise assessment is not feasible without a calculation of the total propulsion system mass and, equally important, a calulation of how much useful thrust is actually developed per expenditure of rocket propellant during each revolution of the rotating arms.

In the area of advanced propulsion, which is our business, whenever we propose a new concept, the first question asked is: how much better is it than the systems we have today? In the case of your unique design, you need to have such a comparison made in order to prove (or disprove) your case.

Thank you for taking the time to share your ideas with us.

Sincerely,

Edward A. Gabris, Director
Propulsion, Power and Energy Division
Office of Aeronautics and Space Technology

March 25, 1987

EDWARD A. GABRIS
DIRECTOR NASA
PROPULSION, POWER AND ENERGY DIVISION
OFFICE OF AERONAUTICS AND SPACE TECHNOLOGY
WASHINGTON, B.C. U.S.A. 20546
ATTENTION RP

Dear Mr. Gabris:

Thank you for your letter of March 12, 1987. In paragraph three of your letter regarding your question on how much?

The answer is,

The test I made years ago on my prototype machine produced a clear and steady thrust ratio of two to one i.e. for every pound of rocket thrust I got two pounds of additional thrust. Whether or not this thrust ratio can be improved upon remains to be seen.

Due to the supposedly lack of interest and no knowledge feedback to me from any source, including NASA, for almost 20 years, I found it impossible to continue this work without support.

As outlined in my letter of September 17, 1986, I separated the two categories namely; space shuttle and space liner. As I visualized my invention for space shuttle, this craft would have an aircraft type take off and landing and would be entirely self contained.

At the present time, I do not visualize an atomic type power source to supply energy to rotate the rocket nozzels. It is my belief that an non atomic aircraft type prime mover can be utilized to power this craft. It is also my belief that direct rocket propulsion is extremely inefficient when utilized as a mass mover from earth to space.

I also believe that in the transportation of human beings into space, safety and reliability is the prime consideration.

As for the deep space liner type space craft, it would have to be assembled in space and it would be powered by three power sources; atomic energy, solar energy and chemical fuel, as this type of space craft would be permanently utilized in space, size and mass would not be limited as in the case of the space shuttle.

I wish to emphasize again, my orbital propulsion test machine has always been available to any organization which would take the trouble to examine it.

/...2

It has been my lifetime belief that often times there is more than one way of overcoming problems in every field of human endeavor.

I also have a concept for a new type of earth to space transportation system that has nothing to do with orbital propulsion, but, would seem to me to be entirely satisfactory for moving non human cargo into space at a very low cost.

I remain,

Yours truly,

George Knapp
GEORGE KNAP ENTERPRISES

National Research Council Canada
Space Division
Ottawa, Canada
K1A 0R6

Conseil national de recherches Canada
Division de l'espace

File Référence 4377-1
Reply Attn: 87/119

30 March 1987

Mr. George Knap
4408 Wildwood Crescent
Burnaby, B.C.
V5G 2M4

Dear Mr. Knap:

I received a copy of your letter, dated 17 September '86 concerning your "Orbital Propulsion Apparatus". I asked a colleague of mine (also an astronaut payload specialist) to review the physics involved in your proposal and I attach a copy of his comments.

In summary, while your system may be made to work, there are more efficient and less complicated methods of achieving vectored propulsion.

Thank you.

Yours sincerely

Marc Garneau

Marc Garneau
Payload Specialist

National Research Council Canada   Conseil national de recherches Canada
Space Division/Astronaut Program Office

FILE NO
DOSSIER NO

DATE 27 Mar. 1987

Reply Attn: 87/116

TO
A
M. Garneau

FROM: B.V. Tryggvason

Re: George Knapp's "Orbital Propulsion Apparatus"

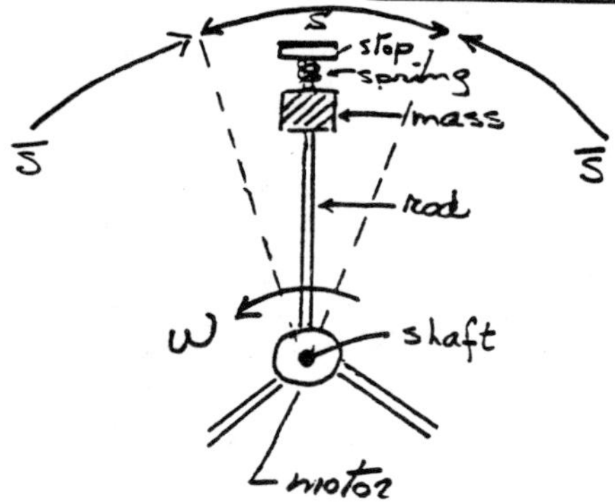

If the mass is moved outward during arc S and the rotational speed maintained by the electric motor, it is true that the force exerted by the stop on the mass towards the centre of rotation increases in proportion to MW^2r. The mass exerts a similar force through the spring onto the stop in an outward direction, which is in turn transmitted to the central shaft through the rod. The net effect is that the central shaft must pull on the rod, stop and mass a bit harder during arc S, and it in turn pulls on the spacecraft a bit harder.

During the remaining arc $\bar{S}$, the mass presumably moves inward again, necessitating a retarding torque from the electric motor to maintain the rotational speed W. To move the mass inward requires a force acting on the mass in addition to MW^2r, which is in turn braced against the stop, which pulls on the rod, which pulls on the shaft, which pulls on the spacecraft. If this is done at 90° beyond arc S, this force would be at right angles to that generated during arc S and can be balanced by a similar counter rotating device on the opposite side of the spacecraft.

It appears then that there can be a net thrust acting in the spacecraft in addition to the direct component of thrust arising from the rocket used to push the mass outward during arc S.

HOWEVER

I find this to be a rather strange device and am not convinced it has much practical merit in comparison to other methods for obtaining thrust. I think also that suggestions made by the author that this can replace current rockets are ill-founded

GEORGE KNAP ENTERPRISES

UNIQUE DESIGNS, CONCEPTS AND INVENTIONS

4408 WILDWOOD CRESCENT, BURNABY, B.C., CANADA V5G 2M4

TELEPHONE (604) 433-7415

April 9, 1987

Marc Garneau
Space Division
National Research Council
Ottawa, ONT K1A OR6

Dear Sir:

Thank you for your letter of March 30, 1987. You will find enclosed, a copy of a letter I recently received from NASA together with my reply.

I believe these two letters should increase your knowledge of "Orbital Propulsion".

I do not think it is fair to me that a representative of the National Research Council would issue a statement without wanting to know all the facts.

For many years I have been carrying this burden alone and at my own expense. Do you not agree that a full open exchange of knowledge is required in order to come to a conclusion based on facts and actual tests, not conjecture. A reply would be appreciated.

Yours sincerely,

George Knap

File Reference - 4377-1
Reply Attn: 87/119

National Research Council Canada — Conseil national de recherches Canada
Space Division — Division de l'espace
Ottawa, Canada
K1A 0R6

File Référence 4377-1
Reply Attn: 87/141

23 April 1987

Mr. George Knap
4408 Wildwood Crescent
Burnaby, B.C.
V5G 2M4

Dear Mr. Knap:

Thank you for your letter dated 9 April 1987. It is obvious that you understand the technical merits of your system better than we do. Accordingly, we would appreciate a detailed engineering analysis which reveals the superior thrust capability of your system (and which accounts for all the factors mentioned in the letter from E.A. Gabris).

Yours sincerely

Marc Garneau

Marc Garneau
Payload Specialist

Canada

Legislative Assembly
Province of British Columbia

OMBUDSMAN

202, 1275 West Sixth Avenue
Vancouver
British Columbia
V6H 1A6
Telephone: (604) 660-1366
Long Distance:
(toll free) 1-800-972-8972
RAPICOM 660-1691

File: 88-52403

September 26, 1988

Mr. George Knap
4408 Wildwood Crescent
Burnaby, B.C.
V5G 2M4

Dear Mr. Knap:

Re: Your complaint against the Expo '86 Corporation

On August 23, 1988 you contacted us regarding the fact that the Expo '86 Corporation had declined your invitation to display and demonstrate your "Orbital Propulsion" invention at Expo '86.

When we met earlier this month, we agreed that this office would not investigate your complaint because such an investigation could not result in a benefit to you in view of the fact that the fair is long since over.

I know that you also discussed with Ms. Kembel of this office the fact that you had raised this matter with Mr. Veitch's office in 1986 and had since sought the return of your file, without success. We contacted Mr. Veitch's office on your behalf and learned that they had gone through an office move since you contacted them and that some files that were considered to contain non-essential information were destroyed at that time. They have no record of your file, but admit that it could have been lost or destroyed in that process. They apologize if that happened, since they did not know that there were documents that you wished to retain.

We do not consider that there is anything further that we can do about your lost file, since we accept that if it was lost or destroyed while in Mr. Veitch's office, that event took place without any intention to cause you harm. They regret any inconvenience to you but there is nothing they can do to restore your file.

In light of the foregoing, we shall not be taking further action on your complaint. The documents which you provided for our inspection at the time of your complaint are enclosed with this letter.

Yours sincerely,

Ron Bohlin
Ombudsman Officer

Encls.

BEACH & BROWN
ATTORNEYS AT LAW
3107 EASTLAKE AVENUE EAST
SEATTLE, WASHINGTON 98102
(206) 325-6789

OBERT W. BEACH
ARD BROWN

PATENTS
TRADEMARKS
COPYRIGHTS

January 27, 1989

Mr. George Knap
4408 Wildwood Cres.
Burnaby, B.C. V5G 2M4
CANADA

Dear Mr. Knap:

I have reviewed the copy of your patent No. 4,087,064 for Orbital Propulsion Apparatus that you sent to me and the other documents.

Your undated introductory note states in the third paragraph,

> I have strong reason to believe that the United States government has a secret program to develop my invention 'Orbital Propulsion', thereby infringing on my patent rights.

Also, the last portion of your paper dated September 17, 1986 states,

> Due to being constantly rebuffed in every direction over a long period of time and having a lot of my mail and overseas cables not being delivered, I have come to some conclusions:
>
> 1) That the Canadian and American Governments are working on my invention in secret thereby conspiring to defraud me of my discovery.

I think that you have done the right thing in presenting your invention to NASA. The reply dated March 12, 1987 does not appear to indicate that NASA would take any action with respect to your invention and you do not explain your

> . . . strong reason to believe that the United States government has a secret program to develop my invention 'Orbital Propulsion' . . . .

Your patent was granted on May 2, 1978 so that it will expire on May 2, 1995. Damages can be collected for infringement of a patent during a period of six years prior to instituting a lawsuit against the infringer. Consequently, I recommend that you watch for any publicity as to developments of the

United States government and Canadian government that would indicate use of your invention before a suit is instituted.

If, "by strong reason" you merely mean "strong suspicion", that would not provide a sufficient basis for instituting a lawsuit against either the United States government or the Canadian government because in such a lawsuit it would be necessary to point to a particular development which you consider to constitute infringement.

Also, a United States patent is infringed by making, using or selling the device of the invention. It is unlikely that NASA or any other government agency would be "selling" your Orbital Propulsion Apparatus so that it would be necessary to rely on your apparatus having been "made and used" by the government. If the government is "developing" your invention as you suspect, it is possible that one has not actually been "made and used" so far.

There is the other aspect that you cannot prevent the United States government from infringing a United States patent but can only collect damages for such infringement. The statute 28 U.S.C. § 1498(a) provides,

> (a) Whenever an invention described in and covered by a patent of the United States is used or manufactured by or for the United States without license of the owner thereof or lawful right to use or manufacture the same, the owner's remedy shall be by action against the United States in the United States Claims Court for the recovery of his reasonable and entire compensation for such use and manufacture.

In order to recover your "reasonable and entire compensation" from the U.S. government it would be necessary to prove the benefit that the government received from the use and manufacture of the invention prior to expiration of your patent. If the government simply experiments with your propulsion system, as distinguished from actually using the propulsion system beneficially on a space craft, for example, it would be highly questionable as to what "compensation" you would be entitled to. The answer to that question must wait until you find out what practical use a government agency has actually made of your invention.

Consequently, based on the correspondence which you sent me, I think it is premature at this time to consider instituting a patent infringement suit against the United States government.

Very truly yours,

Robert W. Beach

Robert W. Beach

RWB:MI

July, 24, 1991

Attention to: Mr. George Knap
Concerning The Orbital Propulsion Spacecraft

The apparatus design in such away, to rotate and release the compress gas from the nozzel in just one of the arms(the arm with the released compress gas is changing with the rotation). And also, there are spring at the end of each arm trying to control the nozzel bar at the end.

For the calculation of centerifical force over each arm, we should make assumption.

Assumption

1)The R.P.M. rotation of the motor is 1800 revolution per minute at the motor pulley.
2)Motor pulley diameter is 3.5 inch.
3)Large drive diameter is 18 inch.
4)The arm length from centre to the nozzel-bar is 20 inch.
5)Weight of nozzel-bar is 10 ounce.
6)Spring constant is 2 pound per 1/16 of inch.
7)The compress gas pressure is 2.5 pound, pushing the nozzel-bar up.

a)Velocity of rotation

Perimeter of the pulley will rotate at 1800 rev/min. Therefore, perimeter of the pulley connected to the motor is 3.14*Diameter= 3.14*3.5= 10.99 inch.

Velocity=Perimeter*R.P.M=10.99*1800/60=329.7 inch per second

b)Radius of rotation

1)For the arm in which compress gas is acting.
By the spring law F=K*X, F is force, K is spring constant, X is displacement.

$$2.5=\frac{2}{1/16} * X \text{ OR } X=\frac{2.5}{32}= 0.08 \text{ inch}$$

2)For the other arm in which isn't compress.
Nozzel-bar is weighted 10 ounce OR is 0.0625*10=0.625 pound
F=K*X OR 0.625=32*X OR X=0.02 inch
3)The compress gas arm = 20 + 0.08 = 20.08 inch.
4)Other arms = 20 + 0.02 = 20.02 inch.

c)Mass calculation

1)Mass of the nozzel-bar with the compress gas.

$$M = \frac{\text{Force exert by the compress gas}}{\text{Gravitational force}} = \frac{2.5}{32.2} = 0.08 \text{ slug}$$

2)Mass of the nozzel-bar with no compress gas.

$$M = \frac{\text{Weight of nozzel-bar}}{\text{Gravitational force}} = \frac{0.625}{32.2} = 0.02 \text{ slug}$$

d)Centerifical force calculation

1)Centerifical force for the nozzel with the compress gas.

$$F = \frac{\text{Mass of the nozzel-bar with the compress gas} * (\text{Velocity of rotation})^2}{\text{Radius of rotation}}$$

$$= \frac{0.08 * (329.7/12)^2}{(20.08/12)} = 36.09 \text{ pound}$$

2)Centerifical force for the nozzel-bar with no compress gas.

$$F = \frac{\text{Mass of the nozzel-bar with no compress gas} * (\text{Velocity of rotation})^2}{\text{Radius of rotation}}$$

$$= \frac{0.02 * (329.7 / 12)^2}{(20.02 / 12)} = 9.05 \text{ pound}$$

Second iteration

Radius of rotation will be changing to:

$F = K * X$ OR $36.09 = \frac{2}{(1/16)} * X$ OR $X = 1.13$ inch

The compress gas arm = 20 + 1.13 = 21.13 inch

$F = K * X$ OR $9.05 = \frac{2}{(1/16)} * X$ OR $X = 0.28$ inch

Other arm = 20 + 0.28 = 20.28 inch

Centerifical force for the nozzel with copress gas.

$$F = \frac{M * V^2}{R} = \frac{0.08 * (329.7 / 12)^2}{(21.13 / 12)} = 34.39 \text{ pound}$$

Centerifical force for the nozzel-bar with no compress gas.

$$F = \frac{M * V^2}{R} = \frac{0.02 * (329.7 / 12)^2}{(20.28 / 12)} = 8.93 \text{ pound}$$

Third iteration

Radius of rotation will be changing to:

$F = K * X$ OR $34.39 = \frac{2}{(1/16)} * X$ OR $X = 1.07$ inch

The comress gas arm = 20 + 1.07 = 21.07 inch

$F = K * X$ OR $8.93 = \frac{2}{(1/16)} * X$ OR $X = 0.28$ inch

Other arm = 20 + 0.28 = 20.28 inch

Centerifical force for the nozzel with compress gas.

$$F = \frac{M * V^2}{R} = \frac{0.08 * (329.7 / 12)^2}{(21.07 / 12)} = 34.39 \text{ pound}$$

Centerifical force for the nozzel-bar with no compress gas.

$$F = \frac{M * V^2}{R} = \frac{0.02 * (329.7 / 12)^2}{(20.28 / 12)} = 8.93 \text{ pound}$$

Since the result of second and third iteration are same. Therefore, no more iteration is needed.

Your truly

F. Fararoom

Fred Fararooni

B.S from University of Illinois 1975 Urbana Champaign

M.S From Bradley University Peoria Illinois 1976

GEORGE KNAP ENTERPRISES

UNIQUE DESIGNS, CONCEPTS AND INVENTIONS

4408 WILDWOOD CRESCENT, BURNABY, B.C., CANADA V5G 2M4

TELEPHONE (604) 433-7415

September 18, 1991

Spar Aerospace Ltd.
Advance Technology Systems Division
1700 Ormont Drive
Weston, Ontario
M9L 2W7

Attention: Mr. Horace Jones

Dear Sir;

On August 27, 1991 I received a letter from Mr. Brian Kerluke of Dynamic Signal Analysis Corporation advising me to get in touch with you regarding a review of the calculations for the "Orbital Propulsion Apparatus."

As I am living on a pension, the funds available to pay you for your work is limited. However, this may give your own people an opportunity to advance their own knowledge in this field. I hereby enclose the information pertinent to the project.

Yours truly,

Mr. G. Knap

Encl.

GK/bab

Spar Aerospace Limited

Advanced Technology Systems Group

1700 Ormont Drive
Weston, Ontario
Canada M9L 2W7
Telephone: (416) 745-9680
Facsimile: (416) 745-4172

SPAR

September 25, 1991

George Knap Enterprises
4408 Wildwood Crescent
Burnaby, B.C.
V5G 2M4

Attention: Mr. G. Knap

Dear Sir:

Thank you for your letter of September 18, 1991 concerning the "Orbital Propulsion Apparatus." Allow me to advise that our field of work does not involve propulsion systems. Further, with regret, I am unable to refer you to a company interested in this specific field.

Please find enclosed with this letter the patent documentation you submitted.

Yours very truly,

A.F.H. Jones
Director
Manufacturing

AFHJ/ef

GEORGE KNAP ENTERPRISES

UNIQUE DESIGNS, CONCEPTS AND INVENTIONS

4408 WILDWOOD CRESCENT, BURNABY, B.C., CANADA V5G 2M4

TELEPHONE (604) 433-7415

Audrey McLaughlin
Leader of the N.D.P. of Canada
Room 531-C Center Block
House of Commons
Ottawa, Ontario
K1A 0A6

September 30,1991

Dear Madam;

As per our telephone conversation of September 30,1991, I formally request that you speak to Svend Robinson, my M.P. in Burnaby, B.C., as to why he would not represent me in the matter of "Orbital Space Propulsion" and my quest for official recognition from the Canadian government.

I believe that there exists a conspiracy between the Canadian and United States governments to deny me recognition for my discovery.

Thanking you very much for your time,
I remain,
yours faithfully,

George Knap

GK/bab

bristol aerospace limited
MAIL: P.O. BOX 874 WINNIPEG, MB., CANADA R3C 2S4
COURIER: 660 BERRY ST., WINNIPEG R3H 0S5

TELEPHONE (204) 775-8331
TELEX 07-57774 / 07-55491
FACSIMILE (204) 885-3195

October 24, 1991

1200-01

George Knap
4408 Wildwood Crescent
Burnaby, BC
V5G 2M4

Subject: Orbital Propulsion Apparatus

Dear Mr. Knap;

Your letter to Joe Tharayil of September 18 was passed to me for review. Your Orbital Propulsion Apparatus concept is interesting but, even if workable, its applications are outside Bristol's sphere of interest.

A complete analysis of the device will need to consider, in detail, all of the transient phenomena which are involved and will necessarily be lengthly and expensive. I suggest that your best course of action is to contact universities; with any luck you can find a professor who would give the problem to one of his students. I'm afraid that I don't have any specific contact that I can suggest, the logical thing is to call the Mechanical Engineering Department of the University of British Columbia.

I wish you luck with your activities.

Yours truly,

BRISTOL AEROSPACE LIMITED

B. L. Jones
Consultant, Defence Systems

/ds

british
columbia
civil liberties association

518 - 119 WEST PENDER, VANCOUVER, B.C. V6B 1S5 - TEL: (604) 687-2919
A Member Of The Canadian Rights And Liberties Federation

30 October 1991

George Knap
4408 Wildwood Crescent
Burnaby, B.C.
V5G 2M4

Dear Mr. Knap,

Thank you for your telephone call this morning, which enabled me to retrieve your file (our computer records do not go back to 1987).

Having reviewed the material you sent to us at that time, I must concur with the evaluation of Mr. Onischak that there does not seem to be any practical assistance we can offer you.

The merits of your proposed invention are obviously beyond our competence to assess. Your accusation that the Canadian and U.S. governments are conspiring to suppress your invention is also beyond our competence, since we have no legal investigative powers.

In sum, it would seem to best way to publicize your claims is the one you have chosen – to write a book detailing your experiences. We wish you every success in this endeavour.

Yours sincerely,

Russell Wodell

Russell Wodell
Administrative Assistant

RW:rw
file: d:\clw\bd\policy\eb14\knap.o91

HOUSE OF COMMONS
CHAMBRE DES COMMUNES
OTTAWA, CANADA
K1A 0A6
(613) 995-7224

AUDREY McLAUGHLIN
LEADER
NEW DEMOCRATIC PARTY

AUDREY McLAUGHLIN
CHEF
NOUVEAU PARTI DÉMOCRATIQUE

Our file A 721

November 8, 1991

Mr. George Knap
George Knap Enterprises
4408 Wildwood Crescent
Burnaby, British Columbia
V5G 2M4

Dear Mr. Knap:

Thank you for sending me copies of your recent letters concerning your "Orbital Space Propulsion" invention. I have forwarded a copy of your correspondence to my colleague Svend Robinson, M.P. (Burnaby-Kingsway).

New Democrats have been urging the government to encourage and recognize the Canadian scientific community and to increase funding for research and development. If Canada is to continue to be a world leader in the area of science we must support our scientists and provide an atmosphere that encourages advances in this area. Unfortunately the Mulroney government has discouraged and weakened science in Canada with its cuts in research and development grants and by its failure to promote Canadian scientists and their inventions.

Once again, thanks for taking the time to write and to share your concerns and invention with me.

Sincerely,

Audrey McL

PASS/APGS